Accounting, Finance, Sustainability, Governance & Fraud: Theory and Application

Series Editor

Kıymet Tunca Çalıyurt, Iktisadi ve Idari Bilimler Fakultes, Trakya University
Balkan Yerleskesi, Edirne, Turkey

This Scopus indexed series acts as a forum for book publications on current research arising from debates about key topics that have emerged from global economic crises during the past several years. The importance of governance and the will to deal with corruption, fraud, and bad practice, are themes featured in volumes published in the series. These topics are not only of concern to businesses and their investors, but also to governments and supranational organizations, such as the United Nations and the European Union. Accounting, Finance, Sustainability, Governance & Fraud: Theory and Application takes on a distinctive perspective to explore crucial issues that currently have little or no coverage. Thus the series integrates both theoretical developments and practical experiences to feature themes that are topical, or are deemed to become topical within a short time. The series welcomes interdisciplinary research covering the topics of accounting, auditing, governance, and fraud.

M. Abu Naser
Editor

New Approaches to CSR, Sustainability and Accountability, Volume IV

Springer

Editor
M. Abu Naser
Accounting and Finance
London Metropolitan University
London, UK

ISSN 2509-7873 ISSN 2509-7881 (electronic)
Accounting, Finance, Sustainability, Governance & Fraud: Theory and Application
ISBN 978-981-16-9498-1 ISBN 978-981-16-9499-8 (eBook)
https://doi.org/10.1007/978-981-16-9499-8

This Springer imprint is published by the registered company Springer Nature Singapore Pte Ltd.
The registered company address is: 152 Beach Road, #21-01/04 Gateway East, Singapore 189721, Singapore

Acknowledgements

As the founding chairman of International Group on Governance Fraud Ethics and CSR (IGonGFESR) and International Women and Business Group (IWBG), I would like to express my gratitude to Dr. M. Abu Naser, who organised and chaired our conferences in 2021 and 2022.

I would also like to thank the rectors and vice-rectors of London Metropolitan University, and the Dean and all academicians of The Guildhall School of Business and Law at London.

Prof. Dr. Kıymet Tunca Çalıyurt
Conference Series
Founder—IConGFESR and IWBC

Contents

Editor and Contributors

About the Editor

Dr. M. Abu Naser is a skilled consultant and researcher with over 17 years of teaching and research experience, with a further seven years of commercial experience in IT. Abu joined London Metropolitan University as a consultant in 2003 and became a lecturer in 2005. He has also taught at the University of London since 2016. Prior to becoming a lecturer, Abu worked as a manager for organisations such as Ford, Granada, Tesco, Command Software Systems, Corona IT, as well as a managing director of a multinational organisation.

Abu has a robust track record of research and use of innovative approaches to teaching and curriculum development. He has taught over 13,000 students at London Met and high-ranked research-based universities in the UK. He taught on and managed various business economics, strategic management, finance, entrepreneurship, leadership and project management modules. He also supervises doctoral students.

He is an active researcher who worked on international collaborative projects and has been published widely in peer-reviewed journals. He also produced sole-authored and edited volumes.

Abu holds interdisciplinary research interests in the areas of accounting, finance and economics, including:

- governance in the global market,
- factors affecting corporate sustainability,
- the implications of organisational activity upon the wider stakeholder environment,
- behaviour in the workplace,
- the relationship between financial and economic measures of performance and corporate culture and strategic development.

Abu is currently working as an editor and reviewer of several academic journals. He has organised, participated, chaired and been a keynote speaker at several academic conferences.

Contributors

Aliu Mulaj Luljeta Faculty of Economics, University of Prishtina "Hasan Prishtina", Prishtina, Kosovo

Beerannavar Chaitra Rangappa School of Law, CHRIST (Deemed to Be University), Bangalore, India

Çalıyurt Kıymet Tunca Faculty of Economics and Administrative Sciences, Trakya University, Edirne, Turkey

Chagas Paulo Andre Matta Universidade Presbiteriana Mackenzie, São Paulo, Brazil

Dedaj But Faculty of Economics, University of Prishtina "Hasan Prishtina", Prishtina, Kosovo

de Souza Vasconcelos Ana Lucia Fontes Universidade Presbiteriana Mackenzie, São Paulo, Brazil

dos Santos Silva Natany Kristine Universidade Presbiteriana Mackenzie, São Paulo, Brazil

Gomes Karine Rosália Felix Praça Federal Rural University of Pernambuco, Pernambuco, Brazil

Jenefa Leena Vel Tech University, Chennai, India

Joseph Christo Director Strategy and Planning, Garden City University, Bangalore, India

Kadyan Jagbir Singh Department of Commerce, Swami Shraddhanand College, University of Delhi, New Delhi, India

Kostanica Fitore Faculty of Economics, University of Prishtina, Prishtina, Kosovo

Leite Emanuel Ferreira University of Pernambuco, Pernambuco, Brazil

Lokanan Mark Data Scientist, Machine Learning, Royal Roads University, Colwood, Canada

Naser M. Abu London Metropolitan University, London, UK

Pallissery Fincy School of Law, CHRIST (Deemed to Be University), Bangalore, India

Prenaj Burim Faculty of Economics, University of Prishtina, Prishtina, Kosovo

Segura Liliane Cristina Universidade Presbiteriana Mackenzie, São Paulo, Brazil

Serbonchini Mariana Alves Universidade Presbiteriana Mackenzie, São Paulo, Brazil

Thomas Freddy School of Law, CHRIST (Deemed to Be University), Bangalore, India

Zaman Mumtahina BRAC University, Mohakhali, Dhaka, Bangladesh

Zeqiri Adelina Faculty of Economics, University of Prishtina, Prishtina, Kosovo

List of Figures

List of Tables

Part I
Introduction

Chapter 1
Overlooking to CSR and Women Issues After 12th IConGFESR and 12th IWBC London 2020 Conference

M. Abu Naser and Kıymet Tunca Çalıyurt

Abstract International Group on Women & Business, Governance, Fraud, Ethics & Social Responsibility started working on women entrepreneurships and governance and CSR-related debate from 2009. Several conferences have been held since the group was founded. This book has been exemplified using papers presented in the 11th International Conference on Governance Fraud Ethics and Corporate Social Responsibility (CSR) and 11th International Women and Business Conference hosted by London Metropolitan University in 2020, in the past conference held in different countries, e.g., 2018 in India; 2017 in Malaysia; 2016 in Canada; 2015 in Brazil. In this book, we have published some of the papers presented at the 2020 conference in London. The debates that take place in the conferences are extensive, outlining the contemporary issues in the world. Our group's fundamental purpose is to recognize, promote, and support excellence in research and to encourage the development and use of knowledge for the benefit of humanity. Following few pages are summarized version of research scholars' contributions from India, Canada, Brazil, Portugal, Turkey, Bangladesh, Estonia, Tunisia, Bosnia, Lebanon, and UK.

1.1 Introduction

It is a fact that technological changes and innovations are transforming the key business processes of universities starting from research and knowledge transfer to teaching. Particularly, e-learning platforms with online access to academic knowledge and learning opportunities are totally shifting expectations of what "bricks-and-mortar" universities will present. There are high pressures to collect first-class students from around the globe and to provide them to evolve in a digital society

M. A. Naser (✉)
London Metropolitan University, Holloway Road, London, UK
e-mail: m.naser1@londonmet.ac.uk

K. T. Çalıyurt
Faculty of Economics and Administrative Sciences, Trakya University, Edirne, Turkey

M. A. Naser (ed.), *New Approaches to CSR, Sustainability and Accountability, Volume IV*, Accounting, Finance, Sustainability, Governance & Fraud: Theory and Application, https://doi.org/10.1007/978-981-16-9499-8_1

via digitally transforming universities (PwC, 2018: 6–7). In addition, digitally transforming universities are supposed to protect excellence in research and teaching standards in an ever more competitive international marketplace. In these hard conditions, digitally transforming universities need to improve the digital capacity of their staff. They need to innovate in not only research and teaching capacity of staff, but in the key business processes and in the practice of being a learning organization continuously (Jones & Goodfellow, 2012: 59–60). The main processes of universities are being revised to adapt digital technologies and to adjust their impacts. This leads to a high pressure for universities that are actively engaged in seeking to establish new approaches of operating that make use of the chances that digital technologies and innovations make available. This situation brings the necessity for digitally transforming universities to rely on systematic performance measurement and evaluation process based on continuous auditing and continuous monitoring (CA/CM) to achieve strategic goals. In this paper, considering the process of digital transformation in higher education sector, i.e., the way digital transforming universities emerge, work, and evolve, the response of the auditors to this challenging digital and more risky audit universe is investigated. First, the digitally transforming university's conceptual framework is explained based on relevant literature. Second, the position of auditors in such a challenging control environment is discussed by explaining the elements of digital and riskier audit universe. As a result of the rapidly changing audit universe in the digitally transforming universities, the areas that need change in the audit culture are explained. In the concluding remarks, policy recommendations are made for stakeholders to achieve value-added audit at the higher education sector based on the facts and future expectations about the digital era. It is a sharing platform independent of the place where people's past knowledge is renewed and new information is taught.

The concept of corporate social responsibility (CSR) is not new to India. For ages, entrepreneurs and businesspersons have tried to help the society through some or other means in its overall development. Various charitable and philanthropic activities were done for the society. The word used for was not CSR, but the spirit was that of CSR. The ancient Indian wisdom of the social distribution of wealth is mainly guided by the "Vedas", which attests that there will be an appropriate conveyance of wealth from the well off to poor people and also censure individuals who appreciate wealth without sharing it with others within the society. With the complete overhaul of erstwhile Company Act 1956 and the introduction of Company Act, 2013, India has witnessed remarkable changes in the society and the corporates operating within the society. Corporates have seriously started assessing and re-assessing their social responsibilities, not only toward their immediate stakeholders but also toward the society in which they have been functioning ever since they were incorporated.

In these challenging times, many governments around the world, including India, have encouraged educational institutions to use technological tools to continue the dispensation of learning to the students. The Government of India through the Ministry of Human Resource Development (MHRD) has created a collection of different platforms for online learning that universities can use during the health crisis (Mhrd.gov.in., 2020). Although the government has set initiatives, the actual usability and usefulness are questionable. India, although having the second-highest internet users in the world only, has 36% internet penetration in the country (India Internet, 2019). Also, there is a significant digital divide seen between the urban and rural sectors. This digital divide in the population proves to be a daunting challenge for many universities to reach every student. This paper tries to identify all these issues by conducting a small qualitative study using secondary data and highlighting some of these challenges.

Generally, e-commerce in developing countries is not studied sufficiently, and there are many aspects of it that need further investigation, especially measuring the intention to continue using e-commerce services from the perspective of post-consumer online purchasing. Additionally, there is a need to study other individual characteristics (Luqman et al., 2016a, 2016b). The same situation is found in Kosovo too. There are a very few studies that deal with consumer online purchasing, even fewer for post-consumer online purchasing, and limited research was conducted to understand the impact of personal innovativeness, e-skills, or perceived security on online services in general. Thus, it is interesting from a research standpoint to explore the main reasons driving e-commerce usage and the individual characteristics effect on continuance intention of e-commerce. Considering this, there is a need to further research this issue by taking different approaches, such as the post-consumer online purchasing approach, and studying different demographics, such as intention for post-consumer online purchasing by students. The use of online purchasing by students is the first phase of accepting online purchasing as a way of buying things via internet. The second phase is continuance usage of online purchasing because online purchasers may frequently show irregular behaviors, meaning that they can first adopt e-commerce and then stop using it for different objective or subjective reasons. This discontinuance of e-commerce usage sometimes relates to the low satisfaction consumers get from the service of an e-commerce provider, and sometimes, despite being satisfied with the service, they discontinue for other personal reasons such as e-skills or personal innovativeness (Hung et al., 2007). With that taken into consideration, it is important to study the impact of individual characteristics that affect the intention to continue using e-commerce.

In higher education, it is observed that interventions of conventional and unconventional education instigate the growth of the desire to undertake among students (Casero et al., 2017). However, Sousa et al. argue that "not only changes in professional occupation or employment have highlighted the last few years, but also changes in the profile of the intellectual patrimony made available, regarding the search for qualification in higher education courses" (p. 91). Such a scenario explains the progressive access to higher education institutions, including those already included in the labor market. Higher education institutions and centers of higher education with entrepreneurial functions are considered favored places for the propagation of these perspectives, in their instructional procedures and investigations (Glavič & Lukman, 2007). The university space by "providing entrepreneurial behavior will collaborate decisively in the disposition of an entrepreneurial culture, driving the forces toward the valorization and encouragement of those who intend to undertake". Thus, it is reasonable to assimilate that the higher education environment has a significant attribution in the understanding of the relationships between teaching entrepreneurship and starting a business (Bae et al., 2014). In the perception of these authors, this is a way to achieve sustainable development through entrepreneurial education and citizen action by enabling, in a way that can include, equally and safely, economic development, social promotion, and environmental preservation.

Gender disparity on the board of Indian IT companies is a continuing saga despite the Indian Companies Act, 2013, mandating at least one woman director in the executive position of public listed companies. Women in India are leaders of varied sectors of businesses and on top leadership positions except in IT companies. This study has found that 14 out of 25 leading IT companies in India have got not more than two women on the board of directors. It has also found that the reasons for said nomination are due to the statutory compulsion to have women on board. Information technology companies are responsible for innovation, business growth, transformation, and diversification. Strategic leadership is the key for IT companies to achieve the above-stated objectives. Inclusiveness in the economic reforms is possible when women are given adequate representation in entrepreneurship and leadership positions in all sectors of industries.

A company's performance essentially depends on the board of directors (Merendino & Melville, 2019), given that they are responsible for approving and overseeing the implementation of strategic goals, the system of governance, and creating company culture (Castellanos et al., 2020). An increasing amount of research finds a strong positive relationship between a firm's sustainability performance and profitability: Companies with high sustainability ratings significantly outperform their counterparts both in terms of stock market value and accounting performance (Birindelli et al., 2018). In other terms, adopting environmental, social, and governance (ESG) best practices leads to a long-term competitive advantage.

For the past few years, there has been an increase in awareness about the corporates regarding helping the stakeholders. This attitude was bolstered when India became the first country to legally mandate corporate social responsibility (CSR) through the Companies Act, 2013. According to the National CSR Portal, 10066 crore was spent in 2014–15 which has increased to 18653 by 2018–19. This shows that the act had a significant impact on the companies in contributing toward CSR in India. Today, with the impact of the COVID-19 pandemic and the lockdown on our public systems and the economy, their responsibility to all stakeholders appears starker than ever, in view of this when the Government of India declared COVID-19 as a notified disaster. The Ministry of Corporate Affairs (vide its General Circular No. 10/2020 dated 23rd March, 2020) was quick to clarify that spending of CSR funds for COVID-19 shall be considered as an eligible CSR activity. Apart from contribution to the PM CARES Fund and State Disaster Management Authority, expenditure incurred on preventive health care and sanitation, ex-gratia to temporary/casual workers over and above daily wages, providing quarantine facilities to those affected, among others, shall be considered as CSR spend. This move was welcomed by the corporate sector as well.

"The advance of technology is based on making it fit so that you don't even notice it, for it is part of everyday life." Bill Gates, Co-founder of Microsoft. In recent years, the developments in the internet and technologies developed new systems accessible to businesses. Social media like online communities is good examples (Smits & Mogos, 2013). The general availability of the internet made it possible for individuals to use social media from Twitter to Facebook and email, achieving a high level of communication without physical meetings (Gruzd et al., 2011). Nowadays, social media is being used in work areas. Many organizations are increasingly interested in adopting technologies for their work operations. The Lebanese banks form samples of those organizations. Because of the Lebanese revolution, Lebanese banks found themselves at risk, and this affected the employees and puts them under pressure. As for the corona epidemic—COVID-19, some people thought that this virus is a normal illness, while others took it seriously and tried hard to protect themselves and their surrounding environment. The common advice to be continuously heard is that citizens ought to "stay home." Social media consequently played an important role in individual personal life and work.

By early 2020, the outbreak appeared to be winding down, with no new cases reported after mid-February. A new coronavirus emerged in China's Hubei Province in late 2019, spreading rapidly to other parts of China and infecting tens of thousands of people. The disease soon traveled to other parts of Asia and the rest of the world. By March, the virus reached more than one hundred countries, leading the WHO to designate the outbreak a pandemic. Many governments impose travel restrictions, quarantines, and lockdowns to try to stop its spread. Iran and Italy become hotspots, with these countries reporting among the highest death tolls outside China. As of late March, there were around 375,000 cases and more than 16,000 deaths globally (CDC.2020).

The fraud triangle (Cressey, 1950), the fraud diamond, and the pentagon fraud model are developed by Crowe (2011). The pentagon of fraud model names five factors that cause fraud: pressure, opportunity, rationalization, capability, and arrogance. The overall objective of the study is to verify the adherence of the fraud pentagon theory to the complaint processes carried out by the United States Security Exchange Commission (SEC) between 2018 and 2019. To meet the purposes of the research, this study has as specific objectives the survey of the files of fraud reporting processes from 2018 to 2019, seeking to identify the dimensions of the fraud pentagon theory; analyze the dimensions present in the situations that preceded the case raised by the SEC, according to its judgment reports; and identify which were the most prominent dimensions in the cases reported by the SEC.

In recent years, various cases of procurement fraud involving Malaysian public sector officers have been disclosed to the public. The activities that led to fraud occurrences among officers in Malaysian public sector agencies include abuse of power in procurement process, lack of understanding in procurement procedures, failure in managing the procurement contract, and others. This study examines the activities that could lead to procurement fraud occurrences from the Malaysian public sector officers' perceptive. The data was collected and gathered through a questionnaire survey approach, which was distributed in 16 ministries that were reported in the Auditor General's report in year of 2015 to 2019. Factor analysis and descriptive statistics were utilized for data analysis. In addition, data reliability was checked by

Cronbach's alpha test, data normality was examined by Skewness and Kurtosis tests, and data validity was tested by using Kaiser-Meyer Olkin test and Bartlett's test. The findings show that 56.02% of the participants perceived that procurement fraud occurred in their working environment. Nevertheless, the occurrence of procurement fraud activities varies according to the job functions. In general, the procurement fraud occurs more frequent within engineering and technical job function group.

Nowadays, policies of countries are seeking on increasing the threshold for the investments in R&D. This way, the technological diffusion is to be spread widely and thus the returns from investments to be shortly. But, the question is whether these policies are generating innovation for the current socio-economic conditions to pursue the returns from R&D? In general, the investments in R&D are causing changes in different fields, as well as they are seen as a tool to increase the standard at the firm's level, by introducing advanced and new products and processes that results to higher income and growth (Bilbao-Osorio & Rodriguez-Pose, 2004). Therefore, due to the technological advancement, firm level is not the only stakeholder that is facing changes. Among the others, the entire economy is transforming. One of the indicators that shows this transformation is the share of employment number and outcome from the economic structure composed of the many activities. Those activities are standardized and are grouped in the three main economic sectors: agriculture, industry, and services (United Nations, 2008). The fluctuation of the employment among the sectors has changed quite a lot from the past decade, by being focused more on the technology-based services—that are developed as result of the innovation. Countries that are referring point for the R&D have their headquarters in the USA—as the biggest investor in the world; in geographic context, they are equally spread in the two other continents. In Europe, they are located mostly in: Germany and the UK and in Asia: China and Japan. As regard to the dispersion of the R&D investments in economic sectors, the recent study conducted with top R&D worldwide performers with 2000 companies shows a close relationship of the investments in ICT sectors. It can be noticed from the Industrial Property (IP) rights that are relatively high in sector of ICT, designs, and less on the trademarks in comparison with other activities. Consequently, R&D investments seem to be headed on the digitalization and technological development worldwide.

Digital transformation in higher education sector is, i.e., the way digital transforming universities emerge, work, and evolve. CSR is a sharing platform independent of the place, where people's past knowledge is renewed and new information is taught. Ancient Indian wisdom of the social distribution of wealth is mainly guided by the "Vedas" which attests that there will be an appropriate conveyance of wealth from the well off to poor people. The concept of corporate social responsibility (CSR) is not new to India. For ages, entrepreneurs and businesspersons have tried to help the society through some or other means in its overall development. Various charitable and philanthropic activities were done for the society. The word used for was not CSR, but the spirit was that. It is a word used by the ancient Indian businesspeople for charitable and charitable activities. It was called CSR.

Corporates in India have started assessing and re-assessing their social responsibilities. Government has created a collection of different platforms for online learning that universities can use during the health crisis. India, although having the second-highest internet users in the world, only has 36% internet penetration in the country. Also, there is a significant digital divide seen between the urban and rural sectors. This digital divide in the population proves to be a daunting challenge for many universities to reach every student.

E-commerce in developing countries is not studied sufficiently and needs further investigation, says study. Use of online purchasing by students is the first phase of accepting online purchasing as a way of buying things via internet. Online purchasers may frequently show irregular behaviors, meaning that they can first adopt e-commerce and then stop using it for different objective or subjective reasons. It is important to study the impact of individual characteristics that affect the intention to continue using e-commerce. In higher education, it is observed that interventions of conventional and unconventional education instigate the growth of the desire to undertake among students. The university space by "providing entrepreneurial behavior will collaborate decisively in the disposition of an entrepreneurial culture," argues Sousa et al. The higher education environment has a significant attribution in the understanding of the relationships between teaching entrepreneurship and starting a business, they argue. The study looked at Kosovo and Albania to study post-consumer online purchasing approach to online purchasing and the intention of students to use online services.

Gender disparity on the board of Indian IT companies is a continuing saga. This is despite the Indian Companies Act, 2013, mandating at least one woman director in the executive position of public listed companies. Women in India are leaders of varied sectors of businesses and on top leadership positions except in IT companies. Then, 14 out of 25 leading IT companies in India have got not more than two women on their board of directors.

India became the first country to legally mandate corporate social responsibility (CSR) through the Companies Act, 2013. According to the National CSR Portal, 10066 crore was spent in 2014–15 which has increased to 18653 by 2018–19. Adopting environmental, social, and governance (ESG) best practices leads to a long-term competitive advantage. Due to the impact of the COVID-19 pandemic and the lockdown on our public systems and the economy, their responsibility to all stakeholders appears starker than ever. The Ministry of Corporate Affairs was quick to clarify that spending of CSR funds for the COVID-19 shall be considered as an eligible CSR activity. The Lebanese banks form samples of those organizations. Because of the Lebanese revolution, Lebanese banks found themselves at risk, and this affected the employees and puts them under pressure. Some people thought that this virus is a normal illness, while others took it seriously and tried hard to protect themselves and their surrounding environment.

By early 2020, the outbreak appeared to be winding down, with no new cases reported after mid-February. A new coronavirus emerged in China's Hubei Province in late 2019, spreading rapidly to other parts of China and infecting tens of thousands of people. Iran and Italy become hotspots, with these countries reporting among the highest death tolls outside China. As of late March, there were around 375,000 cases and more than 16,000 deaths globally (CDC.2020). The fraud triangle (Cressey,

1950), the fraud diamond, and the pentagon fraud model developed by Crowe (2011) are the factors that cause fraud.

In recent years, various cases of procurement fraud involving Malaysian public sector officers have been disclosed to the public. The activities that led to fraud occurrences include abuse of power in procurement process, lack of understanding in procurement procedures, failure in managing the procurement contract, and others. The findings show that 56.02% of the participants perceived that procurement fraud occurred in their working environment. In general, the procurement fraud occurs more frequent within engineering and technical job function group. Nowadays, policies of countries are seeking on increasing the threshold for the investments in R&D. But, the question is whether these policies are generating innovation for the current socio-economic conditions to pursue the returns from R&D? It is seen as a tool to increase the standard at the firm level, by introducing advanced and new products and processes that results in higher income and growth, says the study. The entire economy is transforming.

R&D investments seem to be headed on the digitalization and technological development worldwide. Countries that are referring point for the R&D have their headquarters in the USA—as the biggest investor in the world. Recent study shows a close relationship of the investments in ICT sectors. Industrial Property (IP) rights are relatively high in sector of ICT.

As editors, we hope to receive some feedback for our continuous development of our research and collaborative research endeavors.

References

Bae, T. J., Qian, S., Miao, C., & Fiet, J. O. (2014). The relationship between business education and business intentions: The meta-analytic review. *Theory and Practice of Entrepreneurship, 38*(2), 217–254.

Bilbao-Osorio, B., & Rodriguez-Pose, A. (Fall 2004). Economic, from R&D to innovation and growth in the EU. *Growth and Change, 35*(4), 434–455.

Birindelli, G., Dell'Atti, S., Iannuzzi, A. P., & Savioli, M. (2018). Composition and activity of the board of directors: Impact on ESG performance in the banking system. *Sustainability, 10*(12), 4699.

Casero, J. C. D., Portillo, A. F., Escobedo, M. C. S., & Mogollón, R. H. (2017). The influence of university context on entrepreneurial intentions. In *Entrepreneurial universities* (pp. 65–81).

Castellanos, J. D., & George, B. (2020). Boardroom leadership: The board of directors as a source of strategic leadership. *Economics and Business Review, 6*(1), 103–119.

Cressey. D. R. (1950). The criminal violation of financial trust. *American Sociological Review, 15*(6), 738–743. December, pp. 1–15 retrieved from www.jstor.org

Crowe, H. (2011). Why the fraud triangle is no longer enough. Crowe LLP. Retrieved September 7th, 2019 from http://aibaus.org/wpcontent/uploads/2011/04/20100922AIBAFraud.pdf

Glavič, P., & Lukman, R. (2007). Review of sustainability terms and their definitions. *Journal of Cleaner Production, 15*(18), 1875–1885.

Gruzd, A., Wellman, B., & Takhteyev, Y. (2011). Imagining Twitter as an imagined community. *American Behavioral Scientist, 55*(10), 1294–1318. https://doi.org/10.1177/0002764211409378

Hung, M. C., Hwang, H. G., & Hsieh, T. C. (2007). An exploratory study on the continuance of mobile commerce: An extended expectation-confirmation model of information system use. *International Journal of Mobile Communications, 5*(4), 409–422. https://doi.org/10.1504/IJMC.2007.012788

India Internet 2019. (2019). *IAMIA*. [online] Available at: https://cms.iamai.in/Content/ResearchPapers/d3654bcc-002f-4fc7-ab39-e1fbeb00005d.pdf [Accessed 22 April 2020].

Jones, C., & Goodfellow, R. (2012). The "Digital university": Discourse, theory and evidence. *International Journal of Learning and Media*, 4(3–4), 60. https://www.researchgate.net/publication/275504610_The_Digital_University_Discourse_Theory_and_Evidence (accessed February 09, 2020).

Luqman, A., Razak, R. C., Ismail, M., & Alwi, M. A. M. (2016a). The influence of individual characteristics in predicting mobile commerce usage activities' continuance intention. *Journal of Entrepreneurship and Business, 4*(2), 54–69. https://doi.org/10.17687/jeb.0402.05

Luqman, A., Razak, R. C., Ismail, M., & Mohd Alwi, M. A. (2016b). Continuance intention of mobile commerce usage activities: Does personal innovativeness matter? *Social Sciences (pakistan), 11*(11), 2817–2827.

Merendino, A., & Melville, R. (2019). The board of directors and firm performance: Empirical evidence from listed companies. *Corporate Governance: The international journal of business in society.*

Mhrd.gov.in. (2020). *Technology enabled learning|Government of India.* [online] Available at: https://mhrd.gov.in/ict-initiatives [Accessed 26 April 2020].

PwC (2018). The 2018 digital university staying relevant in the digital age. PWC. https://www.pwc.co.uk/assets/pdf/the-2018-digital-university-staying-relevant-in-the-digital-age.pdf

Smits, M., & Mogos, S. (2013). The impact of social media on business performance. Retrieved from https://aisel.aisnet.org/cgi/viewcontent.cgi?article=1348&context=ecis2013_cr

United Nations. (2008). *International standard industrial classification of all economic activities, revision 4.* United Nations Publication.

Part II
Sustainability in Digital Transformation and COVID Times

Chapter 2
Challenges of COVID-19 Lockdown on Indian Universities

Christo Joseph

Abstract In the wake of the growing coronavirus crisis, 25th March 2020 saw a pan-India lockdown being strictly enforced across the country, initially for a three-week period—which was further extended several times, either uniformly nationwide or on a state-by-state basis. As was expected, this led to tremendous hardship for a large swathe of the country's population. Amid all this, the education sector in India was unduly affected in several ways. Now, nearly a year on, there is still no clarity on when educational institutions will be able to open and function "normally". This paper delves into various aspects related to the way COVID-19 lockdowns affected the education sector in India, universities especially. Offering a multiplicity of perspectives, the objective of the paper is to provide a balanced viewpoint on the way several dynamics of the COVID-19 lockdown in India panned out, especially the challenges that it presented to Indian universities. Following the methodology of interpretivism research philosophy, the paper relies primarily on secondary—yet credible sources of data for providing its perspectives. These especially include a multiplicity of online news sites and other publications that have been gleaned over extensively before providing pertinent information. Additionally, social media sites have also been tapped widely for this paper. While perspectives on such sites are frequently opinionated and subjective, a plethora of them when taken in tandem work towards providing a holistic set of beliefs centred around the primary theme of this paper. For data analysis, a thematic analysis has been used for identifying consistent themes gleaned from secondary data that has been obtained as described above. The findings of the paper point towards significant challenges faced by all stakeholders in Indian universities, with no clear resolution in sight, in the immediate future. As the ongoing vaccination drive intensifies, the challenges faced by these universities are likely to resolve gradually, over a prolonged period.

C. Joseph (✉)
Director Strategy and Planning, Garden City University, Bangalore, India
e-mail: christo.joseph@gardencity.university

M. A. Naser (ed.), *New Approaches to CSR, Sustainability and Accountability, Volume IV*, Accounting, Finance, Sustainability, Governance & Fraud: Theory and Application, https://doi.org/10.1007/978-981-16-9499-8_2

2.1 Introduction

On 24th March 2020, Prime Minister Modi came live on television and started his speech by saying "My dear fellow citizens, Today, once again, I am in your midst to discuss Corona Global Pandemic" (Modi, 2020). Corona Global Pandemic or COVID-19 "is an infectious disease caused by a newly discovered coronavirus" (Who.int., 2020b). The virus is said to have originated from Wuhan, China (Who.int., 2020a). It has spread all over the world, causing countries to impose complete lockdown and deploy social distancing measures to save the uninfected population. India, with its population of 1.3 billion people, was one of the first countries to impose a complete lockdown in the entire region. Additionally, individual states in India also imposed Section 144 CrPC[1] (Standard, 2020), which prohibits the gathering of four or more people in any public areas.

When Prime Minister Modi imposed a complete lockdown from the 25th March to 14th April and further extending it till 3rd May India faced unprecedented challenges with its economy, the healthcare system and most importantly the wellbeing of 1.3 billion people (Goyal et al., 2020).

The Indian higher education sector, which is the "third largest higher education system in the world, next only to China and the United States" (Siteresources.worldbank.org, 2020), has also faced significant challenges due to the lockdown measures taken by the government to contain the spread of COVID-19 virus. This paper focuses on putting forth these challenges affecting the higher education sector (universities) during this time of crisis.

The UNESCO has published on its website that due to the actions of most governments around the world to close educational institutions to contain the virus temporarily, "91% of the world's student population is impacted" and in India, 34,337,594 students are affected at the tertiary level. (UNESCO, 2020). The UNESCO Director-General Audrey Azoulay also commented: "Never before have we witnessed educational disruption on such a scale".

In these challenging times, many governments around the world, including India, have encouraged educational institutions to use technological tools to continue the dispensation of learning to the students. The Government of India through the Ministry of Human Resource Development (MHRD) has created a collection of different platforms for online learning that universities can use during the health crisis (Mhrd.gov.in., 2020). Although the government has set initiatives, the actual usability and usefulness are questionable.

India, although having the second-highest internet users in the world only has 36% internet penetration in the country (IAMIA, 2019). Also, there is a significant digital divide seen between the urban and rural sectors. This digital divide in the population proves to be a daunting challenge for many universities to reach every student. This paper tries to identify all these issues by conducting a small qualitative study using secondary data and highlighting some of these challenges.

---

[1] Criminal Procedure Code (**CrPC**) is the legislation on procedure for administration of substantive criminal law in India.

2.2 Methodology

2.2.1 Research Philosophy

The research philosophy refers to a belief and discusses how the data is to be collected, used and analysed in the research paper (Almalki, 2016). There are several types of research philosophies available which includes positivism, interpretivism, pragmatism and realistic philosophies. The positivism research philosophy includes looking at the world from an objective way. Therefore, the researcher is required to be analytical, and the objective, therefore, is to disassociate himself from personal works and values. Just the opposite of the positivism is the interpretivism research philosophy. The interpretivism research philosophy includes the belief that it is not easy to understand the world based on some principles (Fletcher, 2017). Therefore, the world needs to be perceived subjectively. The subjective approach deals with the experiences of people and the idealism that "the world exists, but different people construe it in very different ways" (Barr Greenfield, 1975).

The pragmatic research depends upon facts and figures. This approach is more result driven. Many authors are of a view that this philosophy does not belong to any particular system. Therefore, the researcher in this philosophy is free to use different techniques and methods. The realistic research philosophy depends upon the interpretivism and positivism principles (Cowan, 2011).

Therefore, keeping the research issue in mind, the interpretivism research philosophy will be used for this research paper. The main reason behind the selection of the interpretivism research philosophy is that it will help in carrying out an in-depth investigation of the research issue and subjectively understand the challenges that are faced. (Cohen et al., 2018)

2.2.2 Research Design

The research design refers to a set of methods and procedures to collect and analyse variables identified in each research issue (Insidehighered.com., 2020). There are different approaches to research design mentioned by (Silverman, 2016) and (Dulock, 1993). The research design is usually of two main types—qualitative research design and quantitative research design. These research designs can be further broken down into five main types of research designs including descriptive design of research, experimental research design, correlational research design, diagnostic research design and the explanatory research design.

For this research paper, a qualitative research approach is taken, and the descriptive research design will be used as it will help the researcher in clearly defining and describing the situation under this study (Koh & Owen, 2000). Moreover, through the data collection, analysis and presentation, the researcher will be able to provide deeper insights about the topic.

2.2.3 Data Collection Method

The data collection method refers to the collection of data from various relevant sources to gather necessary information related to the research issue, for testing the hypothesis and also to evaluate the outcomes. The methods of data collection can be classified as primary data collection methods and secondary data collection methods. This research paper uses secondary data collection method which is "every dataset not obtained by the author", or "the analysis of data gathered by someone else" (Eiras, 2011). The dataset can be sources such as newspaper articles, books, magazines, online portals, also journals.

For this study, secondary data is collected by scraping through online news articles from reputed news sources and the posts/comments on social media sites. Indian university websites and government websites under the Ministry of Human Resource Development (MHRD) Government of India are referred. Social media sites like Facebook and Twitter comments relating to COVID-19 and higher education are segregated and used for analysis in this paper. More than 100 Facebook comments and tweets combined are being collected (Appendix 1).

2.2.4 Data Analysis

To analyse the secondary data collected for this study, a thematic analysis will be used for identifying the themes including the ideas, patterns and topics that keep arising repeatedly. The themes that have been identified from the secondary data sources include the future challenges or the challenges related with uncertainties of future, the personal challenges, infrastructural challenges, technological challenges and the financial challenges (Appendix 1).

2.3 Future Challenges

The future challenges include issues that are related to the uncertainty of the future, lockdown period and its consequences (Nanda, 2020). The future challenges include the impact of COVID-19 on the primary and secondary education in India—the exams of the students have been cancelled or postponed until further notice because of the uncertainty relating to the time frame of the lockdown period. The students are being promoted without any condition or based on certain internal evaluations which may not be accurate. This can potentially have specific negative future impacts because it will affect the education level of the students (Poovanna, 2020).

If we look at students already enrolled in various university courses, both within India and abroad, there are umpteen challenges regarding the way they would go about completing their education towards eventual attainment of the degrees/diplomas in question. Many students had planned out future courses of action, based on completing their ongoing educational pursuits within a predesignated timeframe. Now, as things stand, those timelines have gone haywire, with a cascading effect on other plans, including personal ones such as getting married, living independently and so on.

Moreover, even after a year—with vaccines already out there, and an initial vaccination phase having commenced already, there is little clarity on the way the education sector will really function in ensuing times. As alluded to above, timelines are where the primary concerns lie; there is simply a lot of uncertainty around when things will really come back to "normalcy" in the truest sense.

2.4 Personal Challenges

Personal challenges include all those issues faced by students within themselves due to the lockdown. As an example, there are several personal problems being faced by Indian students who are studying in foreign universities because they have to attend classes at night. As a result, they are compelled to stay up the whole night, and that is affecting their concentration in their classes and causing distress.

Further, many students relied on university hostels for accommodation which also created an idyllic environment for learning. With universities shut indefinitely—and students left without the provision to stay in their hostels, most have been forced to return home, while some (especially outstation students) have managed to avail accommodation in the same city or in the vicinity, invariably having to incur significant additional expenses.

In the case of hostels—or even campus life in general (irrespective of staying in university hostels), students built a rapport and camaraderie between themselves which is extremely hard to replicate on a digital platform. Essentially, we are looking at the "personal touch" being missing, whether it is of friends or other close relations formed on campus, as well as interactions between students and staff members, that are just not taking place since universities remain closed (not necessarily due to the lockdown directly, but the fact that these institutions cannot be opened feasibly in the current circumstances).

In this instance too, the time factor does come into play. Students at the final stages of their university courses in particular—who should have commenced with their professional lives by now, are left with uncertainty around when they could do so. Often, there are societal and financial compulsions (more on that in ensuing sections) wherein starting with a job or at least having a source of income is considered the right way forward. Yet, without a formal degree, that may not be feasible. As a result, while many remain unemployed when they should not be, some others are making do with jobs or some other forms of employment which are tangentially different to what their qualifications merit. An apt instance would be qualified engineers or lawyers, among others, having to take up farming in their rural communities.

Further, the lack of sound mental health is a phenomenal challenge being faced not only by students in Indian universities but also a lot of other stakeholders. If we look at private universities in the country as a prime example, they are not looking at enrolments in the same manner—or to the same extent, as they would in "normal" times. This is causing immense hardship to owners and management of such institutions. Further, a slew of professionals in the higher education sector especially private establishments have lost their jobs. This is, of course, causing immense mental distress.

Coming to the students themselves, there is mental hardship on an unprecedented level. Multiple sources of mental strain are having an adverse impact on the way these students would normally function and go about their day-to-day lives. The usual pattern of life which students would follow has gone completely haywire.

Intricately connected to the infrastructural challenges aspect that we mention in the ensuing point below, it would perhaps be appropriate to touch upon the aspect of "soft skills" here. A campus environment provides an idyllic scenario for students to develop soft skills such as sound interpersonal communication, among others. With regular campus life not taking place, this is, in turn, having an adverse effect on the way students would develop such soft skills.

Skills like these are considered integral to overall success both at a personal as well as a professional level. In fact, once we imbibe such skills, they tend to remain with us for the rest of our lives. Therefore, the absence of a regular campus environment for learning undoubtedly presents significant personal challenges on this front, where holistic personality development is being dramatically compromised.

Finally, a significant personal challenge that manifests itself in the context of virtual or digital learning is the absence of self-discipline observed in several students. Otherwise, in a regular classroom environment, students were compelled to go through the rigmarole of requisite learning processes. Herein, they rely excessively on self-discipline. As we mentioned above, many students simply do not have this level of self-discipline which has now been thrust upon them in the current circumstances. Till we see a resolution on this front where regular classroom sessions restart, such personal challenges are likely to continue manifesting themselves.

2.5 Infrastructural Challenges

Infrastructural issues refer to the discomfort being faced by students due to their inability to visit a place physically (Seethalakshmi, 2020). Moreover, attending online classes becomes more difficult when subjects like law are involved because online learning is not as effective as offline classes. It is noticed by the many comments made on Facebook. One comment read "***Online courses will help in easy access but Not all courses can be made online. Some need direct live contact between the teacher and the student. There should be a balance depending on the course***".—***says Anup Yadav*** a student.

Think about practical lessons in this context. Let us say chemistry laboratory assignments as an apt example; due to infrastructural challenges, these simply cannot be replicated at home—nor can they be imbibed simply by learning about them theoretically. A practical approach to learning on this front is quintessential.

Also think about access to libraries; notwithstanding all the digital learning which takes place today, access to paper publications remains—and will likely continue to be, a quintessential part of the entire learning process. As things stand now (viz. January 2021), this is a significant challenge faced in Indian universities, with limited to no access to libraries.

Infrastructure of this kind simply cannot be replicated, either entirely or at least not so easily. When we combine facilities such as libraries and laboratories—in line with the instances we cited above, with learning that takes place in a physical classroom environment, the holistic development of a student which takes place in these instances cannot be equated with digital forms of learning, especially if the latter is the *only* way forward.

2.6 Technological Challenges

Technological barrier refers to all those barriers that are caused by hardware or software issues that make it difficult to carry out a particular action. Moreover, the internet connection and electricity connection at many places are unstable and that causes a technical barrier to the online classes. As mentioned in the introduction, India faces a huge problem with the digital divide between the urban and rural areas. Online learning becomes exceedingly difficult without the required internet connectivity. One student writes to his teacher "***I came home to Sikkim 10 days ago and we started online classes immediately. But the internet connectivity is so poor that I can barely hear what my professors are saying***", says Urmila H a student.

Moreover, for many students, the whole aspect of learning this way is all too new. In fact, this has triggered a whole new learning curve of its own, where students are first having to go through the process of "learning how to learn"! Additionally, quite a few of them are just not comfortable with learning in a digital environment. Even if the infrastructure is somehow in place, at a personal level, there are hesitations and challenges galore.

To give you an example, many Indian homes are designed in such a way that they do not quite offer a dedicated quiet place to study without distractions. Given the ensuing commotion and lack of privacy, students find it difficult to go about their classes with ease.

Besides the connectivity aspect, also think of technological challenges at a more personal level. For instance, many students lack laptops or desktop computers on which they can easily access online courses, while also interacting with their faculty and course mates. Even in cases where they do, these machines are not powerful enough to allow these students to get online and then go through the desired steps for connecting to online courses and faculty.

While today's new age smartphones have proven to be a fairly feasible alternative to full-fledged computers, they cannot be an outright replacement to the latter. For many processes, access to a computer is essential.

Related to this is the financial aspect (more in the ensuing section dedicated to it). Even when students or their family members know that they have technological challenges, they may not be able to do something about it easily since they are monetarily cash strapped. At the lower end of Indian society, even the most inexpensive of gadgets often proves out of reach for most. So, going beyond computers or smartphones, even if we consider a basic calculator for example, that too may not make a financially viable purchase.

Such technological challenges abound in a society as vast as India. Remember that many students in Indian universities, especially government ones, rely on grants and subsidies. If we take even the most prestigious colleges in India, let us say ones like St. Stephen's or Lady Shri Ram (LSR) or Hansraj College in Delhi University, there are students galore who make it there, thanks to grants as well as special provisions like quotas for those excelling in sports or some such field. In other words, a vast majority—no matter how esteemed the institution in question may be, simply do not possess reasonable financial prowess.

Often, students in private colleges and universities—where such technological bottlenecks typically do not present themselves, are presumed to be representative of students in general, right across the country. This is simply not the case; the deeper we go into Indian society, the more "cracks" appear regarding a gross digital divide, as well as the presence of umpteen technological hardships.

2.7 Financial Challenges

The financial challenges include situations where stress is being caused to different stakeholders due to monetary issues. Various B schools and other institutions are facing problems in the recruitment of new students. The placements are being affected with final-year students losing their job offers. Research students have started losing their research funds which affect their academic progress. One Tweet reads "***Kind request, can a priority be made for giving fellowships, including those projects for which funds have not been sent to respective institutes. I don't know how? But pls help these students if you don't provide in this pandemic situation. It's SHAME***"—Vikas Ghattargi a research scholar.

As was pointed out in a previous section, many university students on the verge of graduating should have been placed and started earning an income by now. This has not happened, causing a cascading array of financial hardship not only for themselves but their families.

In this context, it would be pertinent to glean over the aspect of student loans as well. While not as big an issue as in the USA for example—where student loans have taken on trillion-dollar proportions, they are significant and growing. This is especially true in case of students in private universities, where the tuition fees are high enough to warrant taking on student loans for covering them.

For many students, there is immense uncertainty right now around when they would be able to commence with their formal professions and thus having much needed income. Yet, there are student loans that need to be repaid. If we consider students enrolled in foreign universities, the problem is manifold, especially since the quantum of loans in these cases is incrementally higher.

To date, there has not been much clarity regarding the way relief might be granted to students who have availed student loans. While there is talk of some level of moratorium being offered, the finer aspects of that are yet to be worked out. In most cases, students or their families are compelled to continue repaying student loans, irrespective of whether the students have begun with their careers or started having any kind of an income.

Adding to these woes is uncertainty around timelines for resolution; notwithstanding the fact that vaccinations for COVID-19 have already begun in India, given the country's vast and widespread population, it is reasonably clear that it will take a long time for much of the country to be vaccinated. Largely, the impression especially for now is that all-round resolution will only come about when vaccination has taken place on a large scale. Till then, students and other stakeholders in Indian universities are likely to continue facing a plethora of financial challenges.

2.8 Ethical Considerations in Research

In every research paper, strict guidelines need to be followed for looking after the ethical issues that may arise. Therefore, research that has been undertaken, including both the primary as well as the secondary data, needs to consider pertinent ethical issues. In preparing for this research paper, best research ethical practices are being considered as given in (Ørngreen & Levinsen, 2017). The secondary data that has been collected is entirely from the public domain. No private conversation or comments/messages have been used for the purpose of this paper. It has also been ensured that no plagiarism is done, with all necessary credits being acknowledged.

Further, there is a clear focus on presenting facts, along with a representation of the circumstances that surround them. Little, if any, anecdotal information has been provided in the paper, which ensures the requisite presence of objectivity in it. At the same time, the very essence or purpose of such papers is to present analytical data based on the information that has been cited. This paper fully endeavours towards that end objective.

2.9 Limitations of the Study

In every research paper, limitations are bound to occur irrespective of how well it was planned. This paper is also not immune to this conundrum. This research paper, although written timely, is entirely dependent upon the secondary data. The sample size of comments is also limited, although it points out the critical challenges the COVID-19 crisis brings to the education sector. Also, it must be acknowledged that there is a lack of sufficient previous study on this topic that has been selected for this research paper (Kumar, 2019). If primary data is also collected in addition to the secondary data, the reliability of the outcome will increase. However, this research paper can be considered as a precursor to a more extensive study which includes primary data to be done in future.

Moreover, it is the very inherent nature of this crisis such that gathering primary data in the current circumstances *physically* is a virtual impossibility, given the compulsion for social distancing. However, it can easily be gathered digitally, using means such as surveys, online interviews. Accordingly, secondary sources of data have been relied upon extensively in this paper. Once circumstances become such that gathering primary data *personally* is rendered viable that could easily be considered as a rejoinder to this study. Therefore, this study can easily be looked at as a precursor to COVID-19 lockdown challenges faced by Indian universities, which other researchers can build upon further.

Further, while considering the limitations of this study, we must specially look at the unprecedented nature of the COVID-19 crisis that has manifested itself most unexpectedly. Accordingly, circumstances around it are also bound to be unparalleled in their own way. When we look at the central theme of this paper regarding the way persistent COVID-19 lockdowns posed a plethora of ongoing challenges to Indian universities, we must keep this in mind. In essence, there could easily be ensuing challenges in future coming the way of Indian universities, which we have not documented in this paper. Limitations of this nature must be kept in mind while assessing the overall quality and comprehensiveness of the paper.

2.10 Conclusion or Scope for Future Research

This research paper identifies five significant challenges that the Indian universities are facing during the COVID-19 lockdown. As it is an ongoing crisis, constant and continual research must be done to identify the problems that are faced by the universities. As mentioned in the limitations section, this paper can be considered as a precursor to a more extensive study for more in-depth research by gathering primary data and quantitative data by directly interviewing the board members, students and faculty of various universities and government educational departments. COVID-19 is a crisis that is affecting people all over the world, and research will help in providing more clarity and ideas for innovation.

Towards that end, this paper serves as an ideal basis of learning for institutes of higher education not only in India but across the world. As experts in this realm have noted frequently, this is not the first pandemic to hit the world nor will it be the last. Yet, in this stage of human evolution, where we are at an unprecedented level of global communication and collaboration, such studies are quintessential to preparing adequately for the future.

Undoubtedly, the pandemic has ended up affecting practically every sector of society right around the world. This paper focuses on the impact it has had on higher education in India, specifically universities in the country. As the findings suggest, while the challenges themselves are plentiful, the onset of vaccinations—whose drive has already begun (as of January 2021), should slowly and steadily bring relief.

References

Almalki, S. (2016). Integrating quantitative and qualitative data in mixed methods research-challenges and benefits. *Journal of Education and Learning, 5*(3), 288–296.

Cohen, L., Manion, L., & Morrison, K. (2018). *Research methods in education* (pp. 7–19). Routledge.

Cowan, J. (2011). Research methods in education–By Louis Cohen et al. *British Journal of Educational Technology, 42*(5), E110–E110.

Dulock, H. L. (1993). Research design: Descriptive research. *Journal of Pediatric Oncology Nursing, 10*(4), 154–157.

Eiras, J. (2011). *Data collection and storage*. Newyork: Nova Science Publishers.

Fletcher, A. J. (2017). Applying critical realism in qualitative research: Methodology meets method. *International Journal of Social Research Methodology, 20*(2), 181–194.

Goyal, M., Seetharaman, G., & Sharma, S. (2020). *Coronavirus: Faced with an unprecedented challenge, how is India faring?*. [online] The Economic Times. Available at: https://economictimes.indiatimes.com/news/politics-and-nation/coronavirus-faced-with-an-unprecedented-challenge-how-is-india-faring/articleshow/74986805.cms?from=mdr [Accessed 26 April 2020].

Greenfield, T. (1975). *Linking behavioral research and administrative science*. [Place of publication not identified].

India Internet 2019. (2019). *IAMIA*. [online] Available at: https://cms.iamai.in/Content/ResearchPapers/d3654bcc-002f-4fc7-ab39-e1fbeb00005d.pdf [Accessed 22 April 2020].

Insidehighered.com. (2020). *Coronavirus live updates archive from April 12–18*. [online] Available at: https://www.insidehighered.com/news/2020/04/02/live-updates-latest-news-coronavirus-and-higher-education [Accessed 26 April 2020].

Koh, E. T., & Owen, W. L. (2000). Descriptive research and qualitative research. In *Introduction to nutrition and health research*. Springer.

Kumar, R. (2019). Research methodology: A step-by-step guide for beginners. Sage Publications Limited.

Mhrd.gov.in. (2020). *Technology enabled learning|Government of India*. [online] Available at: https://mhrd.gov.in/ict-initiatives [Accessed 26 April 2020].

Modi, N. (2020). *From midnight, entire country shall go under complete lockdown: PM Modi*. [online] Available at: https://www.narendramodi.in/prime-minister-narendra-modi-says-government-fully-vigilant-on-situation-due-to-covid-19-548761 [Accessed 22 April 2020].

Nanda, P. (2020). *Covid-19: B-schools brace for a rough campus recruitment year.* [online] Livemint. Available at: https://www.livemint.com/news/india/covid-19-b-schools-brace-for-a-rough-campus-recruitment-year-11585746736551.html [Accessed 26 April 2020].
Ørngreen, R., & Levinsen, K. (2017). Workshops as a research methodology. *Electronic Journal of E-Learning, 15*(1), 70–81.
Poovanna, S. (2020). *Covid-19 impact: Karnataka cancels exams for classes 7–9.* [online] Livemint.
Seethalakshmi, S. (2020). *Covid-19 impact: As institutes shut, classes go online.* [online] Livemint. Available at: https://www.livemint.com/news/india/covid-19-impact-as-institutes-shut-classes-go-online-11585603587594.html [Accessed 26 April 2020].
Silverman, D. ed. (2016). *Qualitative research.* Sage.
Siteresources.worldbank.org. (2020). *World bank.* [online] Available at: http://siteresources.worldbank.org/EDUCATION/Resources/278200-1121703274255/1439264-1193249163062/India_CountrySummary.pdf [Accessed 26 April 2020].
Standard, B. (2020). *What is section 144|section 144 Curfew breaking news|section 144 rules.* [online] business standard. Available at: https://www.business-standard.com/about/what-is-section-144 [Accessed 26 April 2020].
UNESCO. (2020). *UNESCO.* [online] Available at: https://en.unesco.org/covid19/educationresponse [Accessed 26 April 2020].
Who.int. (2020a). *Coronavirus disease (COVID-19).* [online] Available at: https://www.who.int/emergencies/diseases/novel-coronavirus-2019/events-as-they-happen [Accessed 26 April 2020a].
Who.int. (2020b). *WHO.* [online] Available at: https://www.who.int/health-topics/coronavirus#tab=tab_1 [Accessed 26 April 2020b].

Chapter 3
Factors Influencing the Continuance Intention to Use e-commerce: An Empirical Study

Burim Prenaj, Fitore Kostanica, Luljeta Aliu Mulaj, and Adelina Zeqiri

Abstract The purpose of this paper is to examine the intention to continue using online purchases among Kosovan consumers. Specifically, the study aims to investigate how personal characteristics, combined with perceived usefulness and satisfaction, affect consumers to continue using online purchasing services. We extend the IT Continuance Model by incorporating personal innovativeness, e-skills, and perceived security/privacy. A survey is conducted with 339 students representing all universities in Kosovo. The structural equation modeling (SEM) technique is applied for data analysis. The results show that all IT Continuance Model hypotheses are confirmed, with satisfaction being the most important factor. Moreover, perceived security and personal innovativeness are found to significantly influence continuance intention, while e-skills do not have a significant impact on it. The main limitation of this study is that the survey only includes university students. Since this group has typically more competent users of new technologies, generalizing the findings of this paper should be done with caution. These findings should serve as good guidelines for targeting youths, but the usage and intention to continue using e-commerce may differ across different age groups. There are a very few studies that deal with consumer online purchasing in Kosovo, even fewer for post-consumer online purchasing. This paper fills this gap by examining the effect of personal characteristics on online purchasing continuance intention. Another contribution is the combination of the IT Continuance Model with personal characteristics in the context of online purchasing services.

Keywords E-commerce · Continuance intention · Satisfaction · Perceived usefulness · Confirmation · Personal innovativeness · E-skills

---

B. Prenaj · F. Kostanica · A. Zeqiri
Faculty of Economics, University of Prishtina, Prishtina, Kosovo

L. Aliu Mulaj (✉)
Faculty of Economics, University of Prishtina "Hasan Prishtina", Prishtina, Kosovo
e-mail: aliu.lule@gmail.com

M. A. Naser (ed.), *New Approaches to CSR, Sustainability and Accountability, Volume IV*, Accounting, Finance, Sustainability, Governance & Fraud: Theory and Application, https://doi.org/10.1007/978-981-16-9499-8_3

3.1 Introduction

The development of technology has changed everyday activities. Hence, innovators and businesses used the chance to benefit from these changes. Motivated by potential profit, they invested in information and communication technologies (ICTs) and used them to develop new products and services with lower costs, increasing sales and profit. However, ICTs do not only benefit businesses, since individual consumers have gained a lot from internet as well. Consumers are making more and more online purchases every day, because they are attracted by the benefits that internet offers in the form of increased convenience and reduced prices (Guzzo et al., 2016). One of the most interesting uses of ICTs is the possibility to buy and sell things online, widely known as electronic commerce or e-commerce. Global sales have increased a lot especially during 2010–2019 and are foreseen to expand by 4.4% in 2020.

Zwass (1996) defines e-commerce as follows: "Electronic commerce (e-commerce) is the sharing of business information, maintaining business relationships, and conducting business transactions by means of telecommunication networks". But this is a rather broad definition, the most common definition used nowadays is a very simple one—it refers to e-commerce as the utilization of internet to buy and sell products and services (Gibbs et al., 2003; Grandon, 2004). In this study, e-commerce refers to the use of all electronic equipment, particularly the use of computers and smartphones, to buy things online.

The flexibility and convenience of e-commerce and its possibility to conduct transactions with great ease make e-commerce to be perceived by the consumers as essential mean to buy things. However, despite the obvious benefits of online purchasing, its use has still not reached satisfactory levels in developing countries. Moreover, the typical adopters in developing countries are educated people with high income (Oluyinka et al., 2013).

Generally, e-commerce in developing countries is not studied sufficiently, and there are many aspects of it that need further investigation, especially measuring the intention to continue using e-commerce services from the perspective of post-consumer online purchasing. Additionally, there is a need to study other individual characteristics (). The same situation is found in Kosovo too. There are a very few studies that deal with consumer online purchasing as a whole, even fewer for post-consumer online purchasing, and limited research was conducted to understand the impact of personal innovativeness, e-skills, or perceived security on online services in general. Thus, it is interesting from a research standpoint to explore the main reasons driving e-commerce usage and the individual characteristics effect on continuance intention of e-commerce. Considering this, there is a need to further research this issue by taking different approaches, such as the post-consumer online purchasing approach, and studying different demographics, such as intention for post-consumer online purchasing by students.

The use of online purchasing by students is the first phase of accepting online purchasing as a way of buying things via internet. The second phase is continuance usage of online purchasing because online purchasers may frequently show irregular behaviors, meaning that they can first adopt e-commerce and then stop using it for different objective or subjective reasons (Lin et al., 2005). This discontinuance of e-commerce usage sometimes relates to the low satisfaction consumers get from the service of an e-commerce provider, and sometimes, despite being satisfied with the service, they discontinue for other personal reasons such as e-skills or personal innovativeness (Hung et al., 2007). With that taken into consideration, it is important to study the impact of individual characteristics that affect the intention to continue using e-commerce.

From what is stated above, this study adopts the IT Continuance Model to explain the intention to continue using e-commerce and further expands this model with several factors in order to better explain this issue. The present study analyzes the factors that influence the intention of Kosovan students to continue using e-commerce, particularly the effect of personal characteristics on intention to continue using e-commerce. The aim of this research is to enhance the knowledge of e-commerce usage in Kosovo, through identification of the most important factors that affect its usage or resistance to use it. Additionally, this research will provide a model that helps to explain the continuance usage of other future systems that help doing business electronically.

This study gives important contribution on clarification of the main factors that affect the intention to continue using e-commerce. In this paper, the intention to continue using e-commerce is explained with the IT Continuance Model proposed by Bhattacherjee (2001), which is adapted to explain e-commerce, but the most important contribution of this research is the extension of IT Continuance Model by adding several other factors in an attempt to increase the explanatory power of the original model. The model was extended by adding individual characteristics, such as e-skills, personal innovativeness, and perceived risk and privacy. Further, the recommendations from this study can be implemented in other developing countries, especially Kosovo's neighboring countries with similar context.

3.2 Theoretical Framework

The framework used in this study is a variation of IT Continuance Model proposed by Bhattacherjee (2001). According to this study, user's continuance decision in this model is the same as the consumers repurchase decision, because both decisions (1) follow an initial (acceptance or purchase) decision, (2) are influenced by the initial use (of IS or product) experience, and (3) can potentially lead to ex post reversal of the initial decision. The continuance model focuses only on post-acceptance variables. According to the original model, IS continuance intention and usage is determined primarily by their satisfaction, which is influenced by perceived usefulness and confirmation.

Since there was a need for broader explanation of the traditional factors influencing intention, in our extension we proposed to include the effects of perceived security and personal innovativeness as factors that impact satisfaction and intention to continue using e-commerce. Further, we included e-skills as a factor that directly determines the continuance intention and use.

The continuance use is more important than initial usage of e-commerce by new consumer (Bhattacherjee et al., 2008). To some models, continuance usage has impact on the consumers adoption decision and intentions (Shaikh & Karjaluoto, 2016). While perceived usefulness on acceptance intention in Technology Acceptance Model (TAM) was strong predictor for continuance intention, satisfaction is the stronger predictor (Bhattacherjee, 2001). The impact on continuance intention for technology readiness is interposed by confirmation, satisfaction, and perception of usefulness (Chen et al., 2013). According to these studies, user satisfaction is not influenced by perceived security and privacy; however, another factor influencing the intention to use internet is the set of e-skills that consumer possesses. This way, e-commerce actually distinguishes users who take benefits from internet. In one way, it is more difficult to change people's attitude and actual use, than changing people's knowledge of media (Van Deursen et al., 2014).

3.2.1 Research Model

This paper uses Bhattacherjee's (2001) "A Post-Acceptance Model of IS Continuance" as a baseline model. It suggests that the intention to continue an IS depends on three factors, namely: the costumer's perceived usefulness (PU), confirmation (CON), and the extent of their satisfaction (SAT). Bhattacherjee's (2001) model further suggests that perceived usefulness affects the continuance intention both directly and indirectly through satisfaction. Moreover, the model suggests that satisfaction serves as a mediator for the effect of confirmation on continuance intention. Following previous studies that have used this model to study consumer's post-adoption behavior regarding e-commerce (Chen et al., 2013; Lu, 2014; Luqman et al., 2016a, 2016b; Susanto et al., 2016), it was deemed appropriate to add three more factors to the original model in order to better fit into our context. Previous studies have found that perceived risk/security plays an important role on consumers behavior toward IS (Centeno, 2004; Libaque-Sáenz et al., 2016; Yousafzai et al., 2003). In our context, the lack of investment for development of local e-commerce services accompanied by the difficulties in shipping and electronic transfer funds has kept the vast majority of consumers inexperienced in online transactions, leading to higher security/risk concerns. In this regard, the experience may also have a direct effect on both perceived usefulness and satisfaction resulting to higher probability that a consumer intends to continue using e-commerce services. Moreover, previous studies have found that individual characteristics and abilities are very important factors in explaining variation in adoption of technologies (Kwon et al., 2007; Luqman et al., 2016a, 2016b; Sun & Jeyaraj, 2013). In this paper, we explore the effects of personal innovativeness and e-skills on Kosovan costumer's intention to continue using e-commerce; we argue that personal innovativeness does not vary much over time;

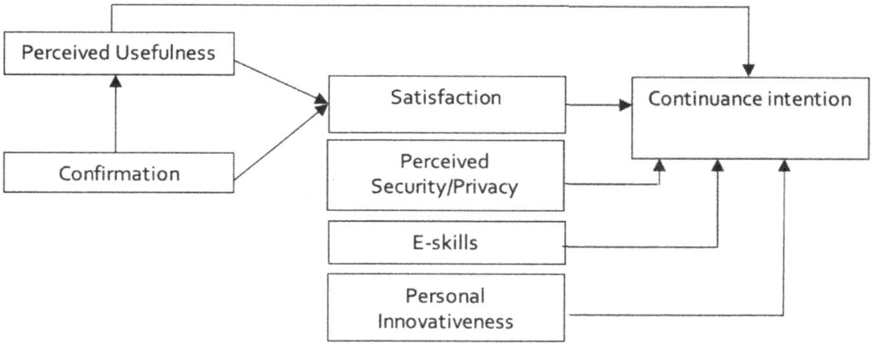

Fig. 3.1 Research model

and it drives the acquisition of e-skills. Therefore, considering our context of underdeveloped e-commerce and difficulties in online transactions, it seems logical that there is massive self-selection into using e-commerce in Kosovo, driven largely by the innate innovativeness and level of e-skills. This model extension suggests that all three of these variables have a direct positive impact on continuance intention (CI), even though these factors may affect or be affected by other factors in the model. In this paper, we will cover only the direct impact of these variables on CI (Fig. 3.1).

3.2.2 *Hypothesis Development*

Continuance intention

According to Bhattacherjee (2001), continuance intention is explained as: the intention to continue using information systems by consumers. Continuance intention is used by Bhattacherjee in the original model of Expectation-Confirmation Theory to understand consumer behavior, consumer satisfaction, and especially behavior of consumers after purchasing. Bhattacherjee emphasizes that continuance intention to use technologies is based on three main factors: satisfaction, confirmation, and perceived usefulness (Alain Yee, 2013). In our extended model, we have included also perceived security/privacy, e-skills, and personal innovativeness, which are expected to positively impact the continuance intention of consumers.

Perceived usefulness

This construct is based on Davis et al. (1989) and Kulviwat et al. (2007); it is used to measure people's belief that technology will affect the productivity and performance of their works. Perceived usefulness according to Brown et al. (2012) was considered one of the strongest determinants that impact use and intention to use technologies. In the model of Expectation-Confirmation Theory, perceived usefulness is considered the most important factor influencing users. Lee (2010) has found that perceived usefulness has the most significant effect on satisfaction. Similarly, Chou et al. (2010) argue that perceived usefulness positively influences continuance intention. Also, Susanto et al., (2016) suggest that perceived usefulness has an important influence in continuance intention. Based on the previous studies, we can propose our research hypotheses:

H1: *Perceived usefulness positively affects continuance intention of e-commerce users.*
H2: *Perceived usefulness positively affects satisfaction of e-commerce users.*

Confirmation

According to Bhattacherjee (2001), confirmation can be defined as users' perception and the compatibility between expectation to use e-commerce and actual usage of e-commerce. In his study, he found that confirmation significantly impacts the perceived usefulness and satisfaction.

Confirmation and perceived usefulness determine the satisfaction of users with information technologies. Thong et al. (2006) found that satisfaction and perceived usefulness were positively affected by confirmation. Lee (2010) showed that confirmation positively affects perceived usefulness. Similarly, the study of Kim et al. (2014) found that confirmation positively affects satisfaction, while Susanto et al., (2016) showed that confirmation affects perceived usefulness as well. Based on these studies, we can propose our research hypotheses related to confirmation:

H3: *Confirmation positively affects perceived usefulness of e-commerce users.*
H4: *Confirmation positively affects satisfaction of e-commerce users.*

Satisfaction

Satisfaction results from customer experience with the product and influences them to buy the product again. So, satisfied customers will repeat the purchase, while dissatisfied customers will no longer use that product (Bhattacherjee, 2001). According to Bhattacherjee et al., satisfaction is an important factor that has significant impact on continuance intention. Satisfaction is a strong determinant that influences continuance intention of using information technologies (Chen et al., 2013). According to Ha and Park (2013), satisfaction is one of the determinants that impacts continuance intention of using a system. Additionally, Luqman et al.,(2016a, 2016b) indicated that satisfaction has a significant relationship with continuance intention. Also, confirmation and perceived usefulness had a significant impact on satisfaction. Based on past studies, we can propose our research hypothesis:

H5: *Satisfaction positively affects continuance intention of e-commerce users.*

Perceived security/privacy

The most important risk factors that impact the use of new technologies are privacy concerns and security concerns (Cheng & Lai, 2012). Dowling and Staelin (1994) defined perceived risk as consumer's perceptions of the uncertainty and adverse consequences of involving in an activity. Pikkarainen et al. (2004) emphasize that security and privacy have an essential role and are the main determinants that impact people to use online services. Furthermore, according to Chen et al. (2013), if the security and privacy are low, users' satisfaction will be decreased. Considering these findings, the proposed research hypothesis is the following:

H6: *Perceived security and privacy positively affects continuance intention of e-commerce users.*

E-Skills

Another important explanatory factor for differences in internet usage by different people is internet skills. Van Deursen and Van Dijk (2009) in their studies have proposed several types of internet skills, including both technical and substantial skills. This study uses a construct of information e-skills built based on the study of Van Deursen et al. (2014). In this regard, the following hypothesis is proposed:

H7: *E-skills positively affect continuance intention of e-commerce users.*

Personal innovativeness

According to Agarwal and Prasad (1998), personal innovativeness is determined as the level at which a buyer is interested in testing any new information technology. Lin and Filieri (2015) and Lu (2014) have established that there are direct and important connections between personal innovativeness and continuance intention. According to Agarwal and Prasad (1998), consumers with higher innovation are likely to adopt an innovation earlier than others, and in order to predict the behavior of consumers or individuals to a novelty, personal innovativeness as a construct needs to be reconsidered. Thus, based on the previous studies conducted, the following hypothesis is proposed:

H8: *Personal innovativeness positively affects continuance intention of e-commerce users.*

3.3 Research Methodology

3.3.1 Research Design

The aim of this study is to examine factors that influence Kosovo's students to continue using e-commerce. In order to do that, this study attempts to test the existing theory about e-commerce continuance intention. So, this study falls within positivist research philosophy, since the research purpose is descriptive of the current usage of e-commerce by students in Kosovo, and it is explanatory because it tries to explain the impact of several factors on e-commerce continuance intention. Therefore, having in mind that this paper attempts to test existing theory about e-commerce continuance intention, a quantitative method was employed, using data collected through a structured questionnaire. Further, the data obtained are analyzed statistically, with aforementioned hypotheses tested using econometric models. The questionnaire for this study is adopted from previous studies in order to ensure validity of the items used (questionnaire items can be found in annex).

Having in mind that the aim was to investigate the Kosovo student's intentions to continue using e-commerce, sample participants were active e-commerce users who had engaged in e-commerce activities in the last year. This was done to exclude students who were not users of e-commerce from answering the questionnaire, which was designed to measure the intention to continue using e-commerce only from customers who already have some experience purchasing via internet. The students were the target population of this study, because young people typically have a greater knowledge of ICTs compared to older people, and the use of technological innovations among young people is higher. This should help us to better study this issue.

3.4 Data

The sample was made of students drawn from both private and public universities in Kosovo, and questionnaires were administered to both undergraduate and graduate students. The data were collected in two forms: through Google Drive, which accounts for two-third of the respondents, and through a face to face survey, which accounts for one-third of the respondents. Before administration of the questionnaire, it was pilot tested with a small group of students and colleagues and was revised after their comments. The final questionnaire is built in four sections. The first one contains personal background questions; the second one contains questions about students' e-skills and personal innovativeness; the third section contains questions about e-commerce usage; and the fourth one contains the constructs used in this paper that are adopted from existing literature. The items on these constructs were measured in a five-point Likert scale, where 1 = strongly disagree and 5 = strongly agree.

A total of 640 students was contacted for this study, of which 301 either did not respond or were not e-commerce users; thus, 339 usable questionnaires were collected in total. The sample of this study is a convenient sample, because it was not possible to find the sampling frame and information about students who use e-commerce in Kosovo. While it was not possible to use a random sample, all the universities and regions of the Kosovo are represented.

3.4.1 Variables of the Study

See Table 3.1.

Table 3.1 Definition of the variables

Variable	Definition	Measures
	Dependent variables	
Continuance intention	*The intention of the respondent to continue using e-commerce activities*	Ordinal variable: 1 = strongly disagree; 2 = disagree; 3 = neutral; 4 = agree; 5 = strongly agree
	Independent variables	
IT continuance model constructs		
Perceived usefulness	*People's belief that e-commerce will affect the productivity and performance of their works*	Ordinal variable: 1 = strongly disagree; 2 = disagree; 3 = neutral; 4 = agree; 5 = strongly agree
Confirmation	*Perception of users for the compatibility between expectation of e-commerce and actual performance*	Ordinal variable: 1 = strongly disagree; 2 = disagree; 3 = neutral; 4 = Agree; 5 = Strongly Agree
Satisfaction	*Respondent satisfaction with e-commerce resulting from evaluation in previous usage*	Ordinal variable: 1 = strongly disagree; 2 = disagree; 3 = neutral; 4 = agree; 5 = strongly agree
Other constructs		
Perceived security/privacy	*Perceived security toward the ability of the e-commerce providers in protecting their information form security threats*	Ordinal variable: 1 = strongly disagree; 2 = disagree; 3 = neutral; 4 = agree; 5 = strongly agree
E-skills	*The ability of the respondents to use internet with competence to fulfill their needs*	Ordinal variable: 1 = strongly disagree; 2 = disagree; 3 = neutral; 4 = agree; 5 = strongly agree
Personal innovativeness	*Propensity of the respondents to experiment with new technologies*	Ordinal variable: 1 = strongly disagree; 2 = disagree; 3 = neutral; 4 = agree; 5 = strongly agree

3.4.2 Sample Characteristics

Table 3.2 summarizes the demographic profile of the respondents such as: gender, study level, and employment. It also shows the results of e-commerce frequency of usage, most used e-commerce providers, and products and services purchased by the respondents.

From a total of 339 respondents, 58.4% were females and 41.6.8% were males. In terms of study level, the majority of respondents were bachelor's students (76.4%), and the second largest group were master's students (20.4%). Regarding the employment status of the respondents, 71.4% were not employed, while only 28.1% of them

Table 3.2 Demographic profile of the respondents and e-commerce usage

Respondent characteristics	Frequency	Percentage
Gender		
Male	141	41.6
Female	198	58.4
Study level		
Bachelor	259	76.4
Master	69	20.4
Ph.D.	11	3.2
Employment		
No	224	71.3
Yes	111	28.1
Frequency of e-commerce usage		
1	38	11.2
2–3	118	34.8
4–6	74	21.8
7–9	41	12.1
10 +	64	18.9
E-commerce providers		
Amazon	91	32.5
eBay	100	35.7
AliExpress	204	72.9
Asos	54	19.3
Products and services bought		
E-books	51	15.0
Software	32	9.4
Digital distribution	21	6.2
Reservations	55	16.2
Clothing and shoes	214	63.1
Technological devices	125	36.9
Jewelry and accessories	147	43.4
Pharmacy	61	18.0
Tickets	53	15.6
Other	16	4.7

Source Own calculations based on survey data

were employed at the time of the survey. Concerning the frequency of usage of e-commerce, the majority of the respondents declared that they use e-commerce activities around 2–6 times a year. The most used e-commerce provider is AliExpress, but respondents declared that they frequently use eBay and Amazon too. Interestingly, e-books and software purchase have a low frequency, which means that students do not view e-commerce as a tool to access their study material. In contrast, clothing and shoes, technological devices, jewelry, and accessories are the most often purchased goods by the respondents.

3.5 Findings

3.5.1 Confirmatory Factor Analysis (CFA)

The present study utilizes structural equation modeling (SEM) technique to analyze the data, while STATA (SEM builder) was used as the tool for the analysis. According to Yang and Lai (2010), when it comes to analyzing causal relationships between latent constructs, SEM offers a good statistical method. In this study, in order to estimate the measurement and structural model, the two-step method proposed by Anderson and Gerbing (1988) is utilized. In the first step, the scale reliability and validity of the model are examined using confirmatory factor analysis (CFA), while the second step focuses on hypotheses testing using path analysis.

3.5.2 Fit Indices of the Model

In order to evaluate if the model used in this study explains the observed data, goodness-of-fit indices of the model are used. Statistics of the fit indices of the measured and structural model are shown in Table 3.3. Four goodness-of-fit measures were used to evaluate the model's overall goodness-of-fit: the ratio of Chi-square to degrees-of-freedom (Chi-square/d.f.), root mean square error of approximation (RMSEA), comparative fit index (CFI), and Tucker–Lewis index (TLI). As Bentler (2006) points out, the lower the Chi-square/d.f. ratio, the better the cut-off value is 3. RMSEA value should be as close to 0 as possible, because 0 indicates that the model fits the data very well, but some authors argue that the cut-off is 0.08 (Hair et al., 2010; Chen et al., 2008). Bentler and Bonett (1980) suggest that CFI and TLI should be as close to 1 as possible, but the cut-off is 0.9.

The results shown in Table 3.3 demonstrate that the measurement and structural model meet the suggested criteria, which means that the model demonstrates an adequate fit and the structural model fits at a moderately acceptable level to the observed data.

Table 3.3 Goodness-of-fit statistics

Fit measure	Criterion	Measurement model	Structural model
$X^2/d.f$	Less than 3.0	2.80	2.76
RMSEA	Less than 0.08	0.078	0.071
CFI	More than 0.90	0.918	0.927
TLI	More than 0.90	0.907	0.919

Chi-square to degrees-of-freedom *($X^2/d.f.$);* root mean square error of approximation (RMSEA); comparative fit index (CFI); standardized root mean square residual (SRMR); fit index (CFI); and Tucker–Lewis index (TLI)
Source Own calculations based on survey data

3.5.3 Measurement Model

The assessment of the accuracy of the model was done through reliability and convergent validity. Cronbach's α and composite reliability (CR) were used to assess the reliability of this study. In order to satisfy the reliability requirements, Cronbach's α coefficient should be at least 0.7 or above. While composite reliability (CR) measures the correlation of the items within the construct, a high CR value means that the factor has high internal constancy. Therefore, the higher the CR value, the higher the reliability (Nunnally & Bernstein, 1994). The results of α and CR for all constructs, as shown in Table 3.4, are between 0.951 and 0.838, therefore, exceeding recommended criteria of 0.7. The CR results are between 0.951 and 0.851, which is greater than the recommended criteria of 0.7. As a result, since all the constructs have a high Cronbach's α and CR, it can be concluded that the questionnaire has high reliability.

Factor loadings and average variance extracted (AVE) were used to assess the convergent validity in this study. To be able to assume that a construct has high convergent validity, factor loadings should be above 0.7, while the recommended criteria for average variances extracted (AVE) are that it should be 0.5 or above (Barclay et al., 1995; Hu et al., 2004; Hair et al., 2011). Table 6 shows that all factor loadings are above the recommended threshold, with the smallest being 0.7 and the largest 0.937. The AVE results are between 0.808 and 0.589; therefore, the convergent validity is ensured as well.

3.5.4 Assessment of Structural Model

Figure 3.2 shows the results of the structural model path analysis, where estimators are presented on standardized coefficients. All but one of the path coefficients are significant, as indicated in Fig. 3.2, therefore, supporting all but e-skills hypotheses. R2 values suggest that the model fits the data adequately and it explains 59.9% of the variance in the endogenous variable perceived usefulness, 74.44% in the variable

Table 3.4 Confirmatory factor analysis results

Construct	Item	Mean	Standard deviation	Cronbach's alpha	Loading	CR	AVE
Perceived usefulness	PU1	3.794	0.9062	0.941	0.869	0.944	0.808
	PU2	3.823	0.9348		0.937		
	PU3	3.766	0.9380		0.88		
	PU4	3.882	0.9424		0.91		
Confirmation	CON1	3.593	0.9052	0.903	0.854	0.906	0.763
	CON2	3.605	0.9072		0.91		
	CON3	3.680	0.9184		0.857		
Satisfaction	SAT1	3.703	0.9003	0.951	0.843	0.951	0.795
	SAT2	3.727	0.9078		0.91		
	SAT3	3.786	0.9171		0.89		
	SAT4	3.777	0.9263		0.924		
	SAT5	3.821	0.9131		0.89		
Perceived security/privacy	PSP1	3.548	0.9420	0.909	0.86	0.914	0.780
	PSP2	3.455	0.9640		0.919		
	PSP3	3.438	0.9935		0.87		
E-skills	ES1	2.430	1.2102	0.878	0.792	0.896	0.634
	ES2	2.145	1.1760		0.846		
	ES3	2.252	1.1841		0.82		
	ES4	2.555	1.2649		0.78		
	ES5	2.722	1.1305		0.74		
Personal innovativeness	PI1	4.040	0.9733	0.838	0.70	0.851	0.589
	PI2	3.217	1.1289		0.81		
	PI3	3.806	1.1206		0.723		
	PI4	3.594	1.0998		0.83		
Continuance intention	CI1	3.667	0.9474	0.916	0.893	0.887	0.725
	CI2	3.461	0.9518		0.82		
	CI3	3.640	0.9548		0.84		

Source Own calculations based on survey data

satisfaction, and 62% of the variance in the continuance intention (as the dependent variable of this study).

Results show that satisfaction has the largest direct influence on continuance intention ($\beta = 0.672$), and it itself is largely impacted by perceived usefulness, which in turn is positively impacted by Confirmation. Perceived security/privacy is the second most important factor impacting e-commerce continuance intention ($\beta = 0.393$), followed by perceived usefulness ($\beta = 0.262$), while personal innovativeness has the smallest statistically significant impact on continuance intention ($\beta = 0.111$).

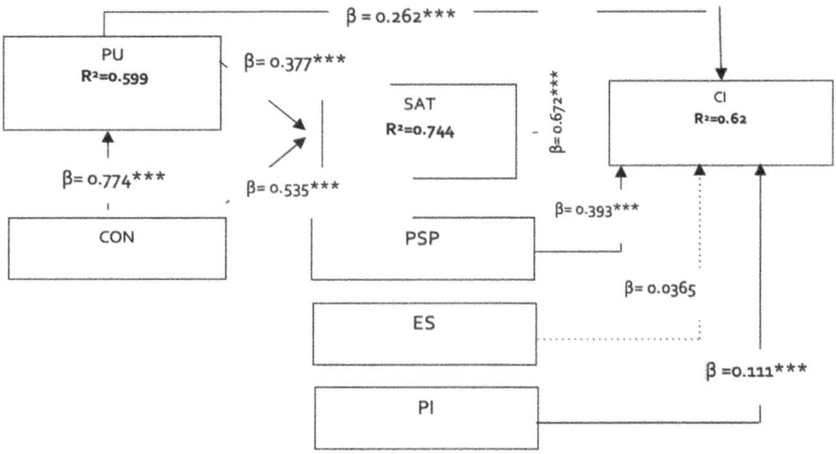

Fig. 3.2 Structural model assessment (standardized path coefficients). *Note* PU—perceived usefulness; CON—confirmation; SAT—satisfaction; PSP—perceived security/privacy; PI—personal innovativeness; ES—e-skills; CI—continuance intention

Finally, e-skills is the only factor that has no effect on e-commerce continuance intention, since it has no statistically significant results.

3.5.5 Discussion

This study proposed and tested a modification of IT Continuance Model, with the aim to test continuance intention to use e-commerce. The path analysis shows that all hypotheses but the e-skills one are confirmed. These results provide enough evidence that the proposed model has good explanatory power and can be used to analyze similar technologies in the future. Table 3.5 shows the results of the hypotheses testing and path analysis, where estimators are reported in standardized coefficients.

As mentioned above, this study used the IT Continuance Model as a basis to explain e-commerce usage intention, and the paths tested from IT Continuance Model are: the effect of confirmation on perceived usefulness and satisfaction; the effect of perceived usefulness on satisfaction and continuance intention; and the effect of satisfaction on continuance intention. All tested paths of the IT Continuance Model are significant. The results of path analysis indicate that satisfaction is the strongest determinant of continuance intention to use e-commerce. Moreover, the results confirm the IT Continuance Model hypotheses that users' confirmation is an important determinant of users' level of satisfaction with a technological system and that confirmation positively impacts users' perceived usefulness. The findings

Table 3.5 Path analysis results and hypotheses testing

Path	Coefficient	Z-value	Testing hypothesis
CON  PU	0.774	28.49***	Supported
R^2	**0.599**		
PU  SAT	0.377	6.46***	Supported
CON  SAT	0.535	9.24***	Supported
R^2	**0.744**		
PU  CI	0.262	3.54***	Supported
SAT  CI	0.672	14.32***	Supported
PSP  CI	0.393	6.68***	Supported
ESCI	0.0365	0.819	**Not supported**
PICI	0.111	2.16**	Supported
R^2	**0.62**		

Note PU—perceived usefulness; CON—confirmation; SAT—satisfaction; PSP—perceived security/privacy; PI—personal innovativeness; ES—e-skills; CI—continuance intention
(*) p-value is less than 0.05; (**) p-value is less than 0.01; (***) p-value is less than 0.001

of this study confirm all the IT Continuance Model hypotheses and are in line with the results of Bhattacherjee (2001); Kim et al. (2014); Susanto et al. (2016); Luqman et al., (2016a, 2016b).

Additionally, this study expanded the IT Continuance Model by adding three more factors (PRP, PI, and ES). Perceived privacy/security has a significant impact on the depended variable—continuance intention. As shown in Table 3.5, the hypothesis that states that perceived security and privacy affects positively the continuance intention of e-commerce is supported. This finding supporting findings from previous research (Cheng & Lai, 2012; Pikkarainen et al., 2004) was expected, because buying things on the internet involves providing sensitive financial and personal information to the e-commerce services provider, and if users have high level of confidence in the e-commerce provider, they engage more easily in e-commerce activities. As expected, personal innovativeness has a positive significant effect on continuance intention; therefore, the hypothesis that PI positively impacts CI is supported too. This confirms the finding of Agarwal and Prasad (1998); Lin and Filieri (2015); and Lu (2014), who state that the individuals who like to experiment with technologies are more likely to use and continue using a technological system. Surprisingly, results of the path analysis show that e-skills do not have statistically significant effects on continuance intention of e-commerce; thus, the hypothesis about e-skills effect is not supported.

3.5.6 Theoretical and Managerial Implications

This study adds to existing knowledge in the use of new technology by specifically improving the understanding of consumer online purchasing and intention to continue using it. An important theoretical contribution of this study is the IT Continuance Model modification to adopt the e-commerce context, which increases the explanatory power of the model. The present study has extended the IT Continuance Model by integrating factors that are important to the context of e-commerce, namely: perceived security and privacy, personal innovativeness, and e-skills. Further, we redefined IT Continuance Model constructs in order to fit the e-commerce context. As shown above, the findings of this study confirm all the IT Continuance Model assumptions. Moreover, among the added factors, perceived security/privacy and personal innovativeness are confirmed to have a statistically significant impact, while e-skills is not.

This study proposes an e-commerce continuance use model, which examines the effect of a set of concrete factors in continuance intention to use e-commerce after having some experience in using these services. E-commerce providers, both international and domestic, when strategizing on how to increase their sales and continuance usage of their services should take into consideration these factors. This model lists a few significant factors that affect the level of user satisfaction and intention to continue using e-commerce, which can help current and future e-commerce providers to increase their sales and continuance intention to use their services. According to the results, satisfaction is a key factor to increase continuance intention to use e-commerce. Therefore, e-commerce providers should take all the steps to improve their consumers' satisfaction, by fulfilling their needs and expectations, making the website user-friendly, increasing website performance, making the delivery of the products fast and accurate, and offering a large variety of products with affordable prices. Further, perceived security and privacy is another important factor on e-commerce usage; therefore, banks should increase e-commerce by guarantying consumers about their privacy through a campaign that offers information about personal information and financial threats protection. If the perception of security is increased, then the likelihood to continue using will also increase; that is because people are worried about providing sensitive financial and personal information on the internet, but if consumers have trust on the e-commerce service provider, they are more likely to use those services.

3.6 Conclusions and Future Research

The purpose of this study was to investigate the most important factors that determine users' e-commerce continuance intention, by using the IT Continuance Model. Although the model has been widely proven in different contexts, there has been little research in examining this issue in Kosovo. Therefore, this study provides evidence on factors that influence Kosovo university students on continuance intention of online purchasing. The study extended the IT Continuance Model by adding additional factors in order to adapt to the e-commerce context. The results confirmed the role of the perceived usefulness, confirmation, and satisfaction on continuance use intention of e-commerce. The findings also highlighted the importance of privacy and security and personal innovativeness, but the effect of e-skills was not statistically significant. Overall, the findings of this paper improve the existing literature on the use of online purchasing by consumers.

Regardless of the fact that the used model was significantly supported, this study has a few limitations that should be taken into account. First, there may be other factors that influence e-commerce continuance usage that have not been included in this study. Second, we studied continuance intention of university students only; therefore, generalizing the results of this study to other groups needs to be done cautiously, since university students are usually more competent users and often the first to adopt new technologies. Consequently, these findings serve as good guidelines for targeting youths, but the usage and intention to continue using e-commerce may differ across different age groups and other demographics. Taking that into consideration, future research could improve on the aforementioned limitations by random sampling and testing the same model across different demographics. Also, adding other factors to this model could further increase its explanatory power.

Appendix: Questionnaire Items

Construct	Item	Table
Perceived usefulness	E-commerce computing may make things more efficient for me as a student (efficiency refers to the time reduced and the chance to quickly find and buy things that students needs compared with other alternatives)	Kulviwat et al., (2007) and Bhattacherjee, (2001)
	Using e-commerce makes shopping activities easier for me	
	I think that e-commerce offers an advantage over other shopping options	
	Overall, E-commerce is useful way to buy things	
Confirmation	My experience with using E-commerce was better than what I expected	Bhattacherjee, (2001)

(continued)

(continued)

Construct	Item	Table
	The service level provided by E-commerce providers was better than what I expected	
	Overall, most of my expectations that I have about e-commerce were correct	
Satisfaction	I find E-commerce to be useful for my shopping needs	Bhattacherjee and Premkumar (2004)
	My choice to use e-commerce was a wise one	
	My experience with using e-commerce was satisfactory	
	I think I did the right thing by deciding to use e-commerce	
	Overall, I was satisfied with the use of e-commerce	
Perceived security and privacy	I think e-commerce service providers have mechanisms to ensure the safe transmission of its users' information	Venkatesh et al. (2011) and Bhattacherjee and Premkumar (2004)
	I feel secure to perform transactions using E-commerce	
	E-commerce is a secure service through which to send sensitive information	
	Overall, e-commerce service is a safe place to transmit sensitive information	
Continuance use intention	I intend to continue using e-Commerce rather than discontinue its use	
	My intentions are to continue using e-commerce than use any alternative means	
	I will frequently use e-commerce services in the future	
Personal Innovativeness	I like to experiment with new technologies	Lu (2013) and Agarwal and Prasad (1998)
	Among my peers, I am usually the first to try out new information technologies	

(continued)

(continued)

Construct	Item	Table
	In general, I would not hesitate to try out new information technologies	
	I would look for ways to experiment with new technologies	
E-skills (information e-skills)	I find it hard to decide what the best keywords are to use for online searches	Deursen et al. (2014)
	I find it hard to find a website I visited before	
	I get tired when looking for information online	
	Sometimes I end up on websites without knowing how I got there	
	I find the way in which many websites are designed confusing	

References

Agarwal, R., & Prasad, J. (1998). A conceptual and operational definition of personal innovativeness in the domain of information technology. *Information Systems Research, 9*(2), 204–2015. https://doi.org/10.1287/isre.9.2.204

Alain Yee-, L. C. (2013). Understanding mobile commerce continuance intentions: An empirical analysis of Chinese consumers. *Journal of Computer Information Systems, 53*(4), 22–30. https://doi.org/10.1080/08874417.2013.11645647

Anderson, J. C., & Gerbing, D. W. (1988). Structural equation modelling in practice: a review and recommended two-step approach. *Psychological Bulletin, 103*(3), 411. https://doi.org/10.1037/0033-2909.103.3.411

Barclay, D., Thompson, R., & dan Higgins, C. (1995). The partial least squares (PLS) approach to causal modelling: Personal computer adoption and use an illustration. *Technology Studies*,. https://doi.org/10.1017/CBO9781107415324.004

Bentler, P. M. (2006). *EQS 6 structural equations program manual*. BMDP Statistic Software.

Bentler, P. M., & Bonett, D. G. (1980). Significance tests and goodness of fit in the analysis of covariance structures. *Psychological Bulletin, 88*(3), 588. https://doi.org/10.1037/0033-2909.88.3.588

Bhattacherjee, A. (2001). Understanding information systems continuance: An expectation-confirmation model. *MIS Quarterly: Management Information Systems*, 351–370. https://doi.org/10.2307/3250921

Bhattacherjee, A., & Premkumar, G. (2004). Understanding changes in belief and attitude toward information technology usage: A theoretical model and longitudinal test. *28*(2), 229–254. https://doi.org/10.2307/25148634

Bhattacherjee, A., Perols, J., & Sanford, C. (2008). Information technology continuance: A theoretic extension and empirical test. *Journal of Computer Information Systems, 49*(1), 17–26. https://doi.org/10.1080/08874417.2008.11645302

Brown, S. A., Venkatesh, V., & Goyal, S. (2012). Expectation confirmation in technology use. *Information Systems Research, 23*(2), 474–487. https://doi.org/10.1287/isre.1110.0357
Centeno, C. (2004). Adoption of Internet services in the acceding and candidate countries, lessons from the Internet banking case. *Telematics and Informatics, 21*(4), 293–315. https://doi.org/10.1016/j.tele.2004.02.001
Chen, S. C., Liu, M. L., & Lin, C. P. (2013). Integrating technology readiness into the expectation-confirmation model: An empirical study of mobile services. *Cyberpsychology, Behavior, and Social Networking, 16*(8), 604–612. https://doi.org/10.1089/cyber.2012.0606
Cheng, F. C., & Lai, W. H. (2012). The impact of cloud computing technology on legal infrastructure within internet—Focusing on the protection of information privacy. In *Procedia Engineering*. https://doi.org/10.1016/j.proeng.2011.12.701
Chou, S. W., Min, H. T., Chang, Y. C., & Lin, C. T. (2010). Understanding continuance intention of knowledge creation using extended expectation-confirmation theory: An empirical study of Taiwan and China online communities. *Behaviour and Information Technology, 29*(6), 557–570. https://doi.org/10.1080/01449290903401986
Davis, F. D., Bagozzi, R. P., & Warshaw, P. R. (1989). User acceptance of computer technology: A comparison of two theoretical models. *Management Science, 35*(8), 982–1003. https://doi.org/10.1287/mnsc.35.8.982
Dowling, G. R., & Staelin, R. (1994). A model of perceived risk and intended risk-handling activity. *Journal of Consumer Research, 21*(1), 119–134. https://doi.org/10.1086/209386
Gibbs, J., Kraemer, K. L., & Dedrick, J. (2003). Environment and policy factors shaping global e-commerce diffusion: A cross-country comparison. *Information Society, 19*(1), 5–18. https://doi.org/10.1080/01972240309472
Grandon, E. (2004). E-commerce adoption: Perceptions of managers/owners of small and medium sized firms in Chile. *Communications of the Association for Information Systems, 13*(1), 8. https://doi.org/10.17705/1cais.01308
Guzzo, T., Ferri, F., & Grifoni, P. (2016). A model of e-commerce adoption (MOCA): Consumer's perceptions and behaviours. *Behaviour and Information Technology, 35*(3), 196–209. https://doi.org/10.1080/0144929X.2015.1132770
Ha, Y. W., & Park, M. C. (2013). Antecedents of customer satisfaction and customer loyalty for emerging devices in the initial market of Korea: An equity framework. *Psychology and Marketing, 30*(8), 676–689. https://doi.org/10.1002/mar.20637
Hair, J. F., Ringle, C. M., & Sarstedt, M. (2011). PLS-SEM: Indeed a silver bullet. *Journal of Marketing Theory and Practice, 19*(2), 139–152. https://doi.org/10.2753/MTP1069-6679190202
Hu, X., Lin, Z., Whinston, A. B., & Zhang, H. (2004). Hope or hype: On the viability of escrow services as trusted third parties in online auction environments. *Information Systems Research, 15*(3), 236–249. https://doi.org/10.1287/isre.1040.0027
Hung, M. C., Hwang, H. G., & Hsieh, T. C. (2007). An exploratory study on the continuance of mobile commerce: An extended expectation-confirmation model of information system use. *International Journal of Mobile Communications, 5*(4), 409–422. https://doi.org/10.1504/IJMC.2007.012788
Kim, G., Ro, H., Hutchinson, J., & Kwun, D. J. (2014). The effect of jay-customer behaviors on employee job stress and job satisfaction. *International Journal of Hospitality and Tourism Administration, 15*(4), 394–416. https://doi.org/10.1080/15256480.2014.961797
Kulviwat, S., Bruner, G. C., Kumar, A., Nasco, S. A., & Clark, T. (2007). Toward a unified theory of consumer acceptance technology. *Psychology and Marketing, 24*(12), 1059–1084. https://doi.org/10.1002/mar.20196
Kwon, O., Choi, K., & Kim, M. (2007). User acceptance of context-aware services: Self-efficacy, user innovativeness and perceived sensitivity on contextual pressure. *Behaviour and Information Technology, 26*(6), 483–498. https://doi.org/10.1080/01449290600709111
Lee, M. C. (2010). Explaining and predicting users' continuance intention toward e-learning: An extension of the expectation-confirmation model. *Computers and Education, 54*(2), 506–516. https://doi.org/10.1016/j.compedu.2009.09.002

Libaque-Sáenz, C. F., Wong, S. F., Chang, Y., Ha, Y. W., & Park, M. C. (2016). Understanding antecedents to perceived information risks: An empirical study of the Korean telecommunications market. *Information Development, 32*(1), 91–106. https://doi.org/10.1177/0266666913516884

Lin, C. S., Wu, S., & Tsai, R. J. (2005). Integrating perceived playfulness into expectation-confirmation model for web portal context. *Information and Management, 42*(5), 683–693. https://doi.org/10.1016/j.im.2004.04.003

Lin, Z., & Filieri, R. (2015). Airline passengers' continuance intention towards online check-in services: The role of personal innovativeness and subjective knowledge. *Transportation Research Part e: Logistics and Transportation Review, 81*(1), 158–168. https://doi.org/10.1016/j.tre.2015.07.001

Lu, L., & Kao, S. F. (2013). The reciprocal relations of pressure, work/family interference, and role satisfaction: Evidence from a longitudinal study in Taiwan. https://doi.org/10.1002/hrm.21532

Lu, J. (2014). Are personal innovativeness and social influence critical to continue with mobile commerce? *Internet Research*. https://doi.org/10.1108/IntR-05-2012-0100

Luqman, A., Razak, R. C., Ismail, M., & Alwi, M. A. M. (2016a). The influence of individual characteristics in predicting mobile commerce usage activities' continuance intention. *Journal of Entrepreneurship and Business, 4*(2), 54–69. https://doi.org/10.17687/jeb.0402.05

Luqman, A., Razak, R. C., Ismail, M., & Mohd Alwi, M. A. (2016b). Continuance intention of mobile commerce usage activities: Does personal innovativeness matter? *Social Sciences (Pakistan), 11*(11), 2817–2827.

Nunnally, J., & Bernstein, I. (1994). *Psychometric theory*, 3rd edn, 1994. McGraw-Hill.

Oluyinka, S., Shamsuddin, A., Wahab, E., Ajagbe, M. A., & Enegbuma, W. I. (2013). A study of electronic commerce adoption factors in Nigeria. *International Journal of Information Systems and Change Management, 6*(4), 293–315. https://doi.org/10.1504/IJISCM.2013.060974

Pikkarainen, T., Pikkarainen, K., Karjaluoto, H., & Pahnila, S. (2004). Consumer acceptance of online banking: An extension of the technology acceptance model. *Internet Research*. https://doi.org/10.1108/10662240410542652

Shaikh, A. A., & Karjaluoto, H. (2016). Mobile banking services continuous usage—case study of Finland. In *Proceedings of the annual Hawaii international conference on system sciences*. https://doi.org/10.1109/HICSS.2016.189

Sun, Y., & Jeyaraj, A. (2013). Information technology adoption and continuance: A longitudinal study of individuals' behavioral intentions. *Information and Management, 50*(7), 457–465. https://doi.org/10.1016/j.im.2013.07.005

Susanto, A., Chang, Y., & Ha, Y. (2016). Determinants of continuance intention to use the smartphone banking services: An extension to the expectation-confirmation model. *Industrial Management and Data Systems*. https://doi.org/10.1108/IMDS-05-2015-0195

Thong, J. Y. L., Hong, S. J., & Tam, K. Y. (2006). The effects of post-adoption beliefs on the expectation-confirmation model for information technology continuance. *International Journal of Human Computer Studies, 64*(9), 799–810. https://doi.org/10.1016/j.ijhcs.2006.05.001

van Deursen, A. J. A. M., Courtois, C., & van Dijk, J. A. G. M. (2014). Internet skills, sources of support, and benefiting from internet use. *International Journal of Human-Computer Interaction, 30*(4), 278–290. https://doi.org/10.1080/10447318.2013.858458

van Deursen, A. J. A. M., & van Dijk, J. A. G. M. (2009). Using the Internet: Skill related problems in users' online behavior. *Interacting with Computers, 21*(5–6), 393–492. https://doi.org/10.1016/j.intcom.2009.06.005

Venkatesh, V., Thong, J. Y., Chan, F. K., Hu, P. J. H., & Brown, S. A. (2011). Extending the two-stage information systems continuance model: Incorporating UTAUT predictors and the role of context. https://doi.org/10.1111/j.1365-2575.2011.00373.x

Yang, H. L., & Lai, C. Y. (2010). Motivations of Wikipedia content contributors. *Computers in Human Behavior, 26*(6), 1377–1383. https://doi.org/10.1016/j.chb.2010.04.011

Yousafzai, S. Y., Pallister, J. G., & Foxall, G. R. (2003). A proposed model of e-trust for electronic banking. *Technovation, 23*(11), 847–860. https://doi.org/10.1016/S0166-4972(03)00130-5

Zwass, V. (1996). Electronic commerce: Structures and issues. *International Journal of Electronic Commerce, 1*(1), 3–23. https://doi.org/10.1080/10864415.1996.11518273

Chapter 4
Knowledge-Based Society: R&D Investments in New Economic Transformation

Luljeta Aliu Mulaj and But Dedaj

Abstract Today's industrial firms are seeking to adopt strategies in which new technologies play key role in the sustainability of the industry and growth. The record on growth, as the most important measure for the long-run success in economy, requires continued innovation in the wide range of products, services as well as advanced methods on the production process and delivery. Furthermore, the importance of the technologic advancement has been confirmed by many studies, especially by emphasizing intellectual human capital and university science on the commercialization of scientific work and discoveries. The study gives an insight on the R&D investments: types, phases, and outcomes from the investments, as well as analysis on the literature for the R&D aspects, innovation, and technological improvement. Besides, evidence is given from the 30 world countries on R&D investments as share of GDP. Comparison among countries are made, as well as among three types of the group composed of 10 countries: (a) countries with the highest GDP in the world, (b) countries with the highest GDP per capita, and (c) Balkan countries. The estimation is done based on the data for year 2017. Additionally, there are analyzed and elaborated policies for investments in R&D at the European level, and evidence is provided also for the world top R&D investments distribution.

Keywords Investments · Research and development · Invention · Innovation · Technological progress

4.1 Introduction

Technology was developed as a result of the human beings needs to defeat the naturally occurring forces that arise during the production process, while humans by their work transforms the work object into the final product. In the low stages of the human society development, human beings in order to fulfill their needs by using their physical strength they have found tools with their commitment for providing

---

L. Aliu Mulaj · B. Dedaj (✉)
Faculty of Economics, University of Prishtina "Hasan Prishtina", Prishtina, Kosovo
e-mail: but.dedaj@uni-pr.edu

M. A. Naser (ed.), *New Approaches to CSR, Sustainability and Accountability, Volume IV*, Accounting, Finance, Sustainability, Governance & Fraud: Theory and Application, https://doi.org/10.1007/978-981-16-9499-8_4

the necessary life products. In this stage of development, humans did not differ from other nature objects. To get out of this situation, where they actually played powerless role of the object of nature processes, changes into the subject and the government of natural changes have been made possible by two factors:

1. High degree of intelligence development, which has not been expressed in other facilities.
2. Ability of the mind to create tools for work, through hands from the acquired and accumulated knowledge about the nature and surroundings. Therefore, these two factors are the major ones, which have influenced the man to slowly begin to produce the tools that are needed in the production process.

The timeframe for the development of the technique involves several decades—centuries. In a retrospective look at technological development, uneven development pace can be distinguished by the intensification of the pace of technological development influenced by the first social division of labor, i.e., the division of agriculture with farming, some 5500 years before the new era. In this stage, agriculture was divided as independent manufacturing branch, which as a result submitted the request for the production of necessary tools for work. Essential changes occur when the production of the machine tools were presented. This way, humans are released from the physical work and more likely to deal with creative work, enabling them to construct more complicated machinery.

It is thought that the first complete machine was the spinning machine that was constructed by John Vatt in 1735, which is estimated to have begun with the industrial revolution.

Technological innovations began to rank one after the other. Between 1804 and 1830, there are some great discoveries in the field of rail transport and automobile and ship constructions. However, application of machine tools at the beginning and steam machine at later stages, even though have contributed to the increase in production; the reduction in costs; and the "promise" of greater profits, which on that time have often faced resistance by the owners of the enterprises. This mostly because the labor force has been relatively cheap, while new investments have had to be made for the application of these new technologies, whereas the effects of these investments have been observed later.

The conviction that technological advancement could affect economic development, most American economists and beyond have been influenced by a scientific work of the author Moses Abramowitz, published in 1956. In his statistical research, he has pointed out that the growth of the USA net product per capita in the last 80 years has been almost linked to something that was neither labor nor capital, and later, he did called "residual" that implies technological advancement in general and affects the efficiency of all resources (Dedaj, 1998).

Although with all the difficulties that have emerged, technological advancement has penetrated production, which has been judged to be an inalienable process. These innovations of new technologies in production have also had a significant impact on human activity. At the beginning, new technologies have facilitated man's work in the production process, where later was removed from direct production affecting it to deal with the other phases of the production process (preparatory and controlling phase). Therefore, humans have exploited their intellectual and non-physical strength, and this way, even have been able to help technological advancement.

The importance of the technologic advancement has been confirmed by many studies, especially by emphasizing intellectual human capital and university science on the commercialization of scientific work and discoveries (Gregorio & Shane, 2003). In terms of the economic and social development, (Audretsch et al., 2009) (Lin, New structural economics: a framework for rethinking development, 2012), positive relationship is shown in economic growth and international trade. The well-established relationship resulted on the increased income per capita and economic growth and continued with industrial upgrading and technological innovation, as a modern phenomenon.

Many authors have worked on the issues related to the scientific researches, technological change, and economic transformation and growth. Among others, Joseph A. Schumpeter, in the theoretical contribution of the theory of economic development, had an insight of the central role technology and innovation plays in economic growth.

Therefore, the spending of the total budget in the science is significant to be analyzed, as it has significant effects on the commercial introduction and economic growth. Additionally, the close relationship of economic growth or stagnation is examined by (Maclaurin, 1953) that has resulted to be more affected by the propensity to invent and innovate in comparison to the development of the pure science, as it rarely leads to an immediate technological change that could be discovery and applicable for wider society. Additionally, scientists, engineers, and R&D facilities are considered as leading engine toward technological advancement especially in the field of aviation, chemical processes and products, and electrical systems and equipment (Nelson, National Innovation Systems. A comparative Analysis, 1993).

Nowadays, policies of countries are seeking on increasing the threshold for the investments in R&D. This way, the technological diffusion is to be spread widely and thus the returns from investments to be shortly. But, the question is whether these policies are generating innovation for the current socio-economic conditions to pursue the returns from R&D? In general, the investments in R&D are causing changes in different fields, as well as they are seen as a tool to increase the standard at the firms level, by introducing advanced and new products and processes that results to higher income and growth (Bilbao-Osorio & Rodriguez-Pose, 2004).

Therefore, due to the technological advancement, firm level is not the only stakeholder that is facing changes. Among the others, the entire economy is transforming. One of the indicators that shows this transformation is the share of employment number and outcome from the economic structure composed of the many activities. Those activities are standardized and are grouped in the three main economic sectors: agriculture, industry, and services (UnitedNations, 2008). The fluctuation of the employment among the sectors has changed quite a lot from the past decade, by being focused more in the technology-based services—that are developed as result of the innovation.

The chapter gives an insight on the R&D investments, how they are divided, phases and outcomes from the investments, as well as analysis on the literature for the R&D aspects, innovation, and technological improvement. Besides, evidence is given from the 30 world countries on R&D investments as share of GDP. Comparison among countries is made, as well as among three types of the group composed of 10 countries: (a) countries with the highest GDP in the world, (b) countries with the highest GDP per capita in the world, and (c) Balkan countries. The estimation is done based on the data for year 2017. By the end of chapter, there are analyzed and elaborated policies for investments in R&D at the European level, and evidence is provided also for the world top R&D investments distribution.

4.2 Literature Review

Today's industrial firms are seeking to adopt strategies in which new technologies play key role in the sustainability of the industry and growth. The record on growth, as the most important measure for the long-run success in economy, requires continued innovation in the wide range of products, services as well as advanced methods on the production process and delivery (Bender & co., 2008); (Audretsch et al., 2009). On the book "Industrial Transformation; Environmental Policy Innovation in the United States and Europe" by (Bruijn & Norberg-Bohm, 2005), authors found out that set of policy innovations is necessary for such an industrial transformation, aiming a sustainable industrial society. In this context, from the inception of the industrial revolution, the evolution of the modern economic system passed changes of the macro-economic magnitude in the absolute levels of, e.g., GDP, investments; employment; consumption, etc., which implies the changes on the composition of their structure–dynamics (Pasinetti, 1993). Lot of efforts are put by the scientists for finding solutions toward positive changes like: on decreasing costs, increasing production, and delivery by increasing quality as well.

Problems identified to have no immediate solution are usually carried out by scientists via scientific researches. Author (Souoza, 2010) explains that "Research is performed using specific criteria—these criteria constitutes to a classical method, widely accepted as the best available procedure to solve problems. This procedure is called the 'scientific method'. Science is about explaining the nature of the problem, not blindly trusting that you know what the problem is". In this context, scientific research at industry and at the university is quite different. This is because in industry, some of the criteria are skipped for the practical reason that can lead to the wrong results. Close to the research activity is also the development. Therefore, term research and development has been brought in 1930s in several annual reports. The use of the term: research and development expresses the importance given by the industrial laboratories on the product development and processes (Godin, 2006). Research and development have the components, where most of the new technologies have by any means been dependent in a new scientific knowledge. This way, technological innovation is realized from the advancement of the scientific theories that in one way explains their design and performance (Mowery & Rodenberg, 1998).

The technological progress is described by (Schumpeter, 1943) as the way, where teams of trained specialists can predict what is required. This process leads to the technical change that is initiated by the private or public activity: Research and development—R&D, the output of which is at the beginning intangible and is known as "knowledge capital". It therefore should be used together or merged with other inputs in order to generate revenues. According to Schumpeter, the process of technological change is divided in three phases:

(a) "Invention of a new product or process.
(b) Innovation, which is the transformation of an invention into a commercial product, accomplished through continual improvement and refinement of the new product or process.
(c) Diffusion, which is the process of gradual adoption of the innovation by other firms or individuals from a small niche community to being in widespread use" (Loschel, 2002).

Linear model of innovation was among the first frameworks developed that aimed to explain science and technology relation in an economy. It postulated that "innovation starts with basic research and is followed by applied research and development, and ends with production and diffusion" (Lin, New structural economics: a framework for rethinking development, 2012).

Based on the component of this process, the technological progress can be divided in four phases that are similar to the Schumpeter division that will be explained in details below (Dedaj, 1998):

1. Fundamental scientific research.
2. Applicative scientific research—invention.
3. Developed scientific researches—innovation (application of the invention).
4. Diffusion (Devetakovic, 1989; Mesaric, 1971).

There are also some other definition. Authors: (Norman & Verganti, 2013) define researches as:

1. Basic researches—that have objective to new meanings, but not defined the usage.
2. Design-driven research—that have intention to apply new meanings in products and other phases.
3. Human-centered research—aims to explore people's meaning assigned to products and then design products that fit to the identified meanings.
4. Tinkering—means when people plays with technology or product without any intention for enhancement of the meaning or practical use.

However, the last phase, tinkering, in some cases can lead to the brilliant results and new products that are completely accidental. Below are explained the four phases as per division of the author (Dedaj, 1998).

The first phase: Fundamental scientific research represents the work that aims to widen the scientific knowledge, where at this stage, it is not of interest to apply this knowledge in practice. Thus, these researches represent the first phase of the technological advancement that includes humans' activity to get deeper understanding of the nature or social phenomenon in general. This phase can be considered as pure science that does not necessary lead to technological change or patentable invention (Maclaurin, 1953). These scientific researches even though do not have the starting point to apply in new technology; these researches represent the main scientific basis for new discoveries. Further, these researches are divided into:

- Free fundamental scientific researches.
- Oriented fundamental scientific research (Dedaj, 1998).

First type of the fundamental scientific researches as objective has to widen the knowledge for natural and society in general, but not a defined one, while the oriented fundamental researches have the objective defined from the initial phase that seeks to realize.

The second phase: Applicative scientific research—invention (idea, model) produces idea, by taking into account the practical application of the scientific results. This stage seeks to find opportunities of the gained knowledge from the previous phase to concretely implement. Inventions itself do not generate innovation, unless are commercialized and diffused, thereby raising output in an economy (Audretsch et al., 2009). Additionally, invention according to some authors is supposed to be without any importance about economic analysis, but it is seen to be an intellectual creativity (Godin, 2006). In this context, invention differs from innovation, where innovator produces idea, while in the business aspect, it serves to the entrepreneurs, who in fact they get things done (Schumpeter, The creative reponse in Economic Histrory, 1947).

Third phase: Development scientific research—innovation. This phase represents scientific research work in which research usually by using the results of the two previous phases and aims to implement in practice (Dedaj, 1998). Thus, research try the invention to implement in a daily life. Innovation is when new and/or improved product, service, or process is introduced commercially from the invention (Maclaurin, 1953). It covers much more areas of the developments than invention. Furthermore, innovation is described to be highly heterogeneous phenomenon (UNIDO, 2016). It is composed of some determinants, out of which could be sartorial innovation; R&D expenditure; market condition; access to finance, skilled workers, and knowledge spill overs. Moreover, innovation has a multiplying effect in economic and social developments through technological change, which is measured through patents that represent kind of codified knowledge generated by the individuals, firms, and organizations (Archibugi & Coco, 2004). The major investments on R&D are made by the governments. The support is crucial also for other segments, such as health industry, where advanced economies to keep the trend of achieving technological innovation must invest continuously in R&D (Lin, New structural economics: a framework for rethinking development, 2012).

Forth Phase—Diffusion: It is the last stage in the evolution of any discovery, a phase that lasts from the moment of its first application (innovation), to its widespread in practical massive use, and is of great importance because at this stage, it verifies the achieved results in the previous phases (Dedaj, 1998). Schumpeterian process of the technical change is composed by three phases of that explains diffusion as acceptance and use of the innovation by the community. According to Schumpeter (1935, 1912), innovation is "defined as:

1. The introduction of a new good or a new type of good in the market
2. The introduction in an industry of a new method of production
3. The opening of a new market in a country
4. The attainment of new sources of supply of raw material or not completely manufactured
5. The implantations of a new organizational structure.

In view of this, different types of innovations can be identified. First, *technical innovations* relate to products and services, technical productive process and service operations. Second, *organizational innovations* relate to the structure and administrative direction of the organization, processes, and human resources. Third, it is possible to distinguish *process innovations,* restricted to the specific company in which it takes place, and the *product innovations* that allow the satisfaction of demand and the expansion of market share of the company. In addition, we can define an innovation as radical if fundamental changes can be realized in the activities of the firm; an innovation is called gradual when those changes are marginal with respect to habitual practices" (Giacchero et al., 2007).

In the context of diffusion, authors (Cantner & Malerba, 2006) state that: "there exist other forms of knowledge diffusion and innovation of higher importance to the relationships between agents than their ability to gather information and knowledge. These relationships between agents and institutions, become an important element in the creation of dynamic competitive advantages, based on the formation, transmission and evolution of knowledge". Innovation is crucial for "production" of new knowledge and exploitation of existing economic knowledge (Rodríguez-Pose & Crescenzi, 2010). Great macro-economic importance is given to the innovation especially by the countries that are highly industrialized, as innovation is considered to be responsible for half of economic growth (Ignat, 2017). Innovation capability is considered to be not only a success factor, but also a necessity for the businesses to keep advantages and prevent from the harsh competition. Hence, companies must increase the range of innovative products and develop them for a shorter period, since from the globalization, the innovation cycles are shorten; investments for innovation are increasing, therefore increasing also the diffusion of innovation. Examples for the shortened period of the innovation cycle are: smartphones, notebooks as from the competition, the development took only some months and not more.

Diffusion of a new technologies and techniques encompasses the elimination of various inefficiencies and taking advantages of the economies of scale as well as it includes the advances in science in general. On the other hand, (Mueller, 1962) diffusion of information is not a cost-free process, because of the patents and secrecy.

About the implementation, countries that have more advanced economies tend to diffuse innovation and to jointly take actions with government and industry (UNIDO, 2016).

4.3 Data Analysis

4.3.1 Expenditures in R&D by the World Economical Leading Countries

Developed countries in general tend to invest more in R&D than the other less developed. From the evidence of more than three decades, the developed countries have mainly increased almost their spending on R&D in each year. In the table below are 10 most developed countries—with largest GDP in the world and their spending in R&D for around three decades (Table 4.1).

From the table, we can see that 3 out of 10 countries have decreased their investments in research and development activities. Country with the largest amount spent in 2016 is Korea Republic with 4.22 % of the GDP, followed by Germany with 2.9 % of the GDP and the USA with 2.7 of the country's GDP. On the other hand, some of the countries have decreased their investments from year 1981 to 2016, e.g., the UK, from investing 2.23% of the country GDP in R&D in 1981, has decreased to 1.64% in 2016, followed by Russian Federation, 1.32% in 1991 to 1.09% in 2016. Canada has invested 1.2 % in 1981, while the other decade has increased the amount of investments to 2.02 in 2001 and again decreased to 1.60 in 2016.

Table 4.1 R&D spending as % of GDP by countries with the largest GDP in the world

No.	Country	1981	1991	2001	2011	2016	+ Increased Investments/ – Decreased investments
1	The USA	2.26566844	2.613973	2.638323	2.769648	2.744179	+
2	China	/	0.724634	0.940331	1.775393	2.108264	+
3	Japan	2.046936901	2.676611	2.97183	3.244766	/	+
4	Germany	2.351612025	2.395721	2.385577	2.795624	2.931682	+
5	The UK	2.238511994	1.863582	1.593739	1.647601	/	–
6	France	1.867444786	2.277471	2.138043	2.191615	2.24805	+
7	Italy	0.83100936	1.14589	1.044907	1.209835	1.285659	+
8	Canada	1.202600777	1.540215	2.028312	1.791323	1.604073	–
9	Russian Federation	/	1.328521	1.093834	1.012609	1.096903	–
10	Korea Republic	/	1.740828	2.341084	3.743613	4.227441	+

Source The World Bank (https://data.worldbank.org/)

4.3.2 *Expenditures in R&D from the Countries with Highest GDP Per Capita*

Countries that are smaller with population, but with largest GDP per capita, have propensity to invest in R&D, as they are quite industrialized, and return from the investments seems to be as tool to keep their living standards as high as they have. In the table below are shown the amount of the expenses in R&D activities as share of GDP of respective country (Table 4.2):

From the data retrieved from world bank, we can see that also the countries that have high GDP per capita keep increasing the investments in R&D. Almost all countries (9 out of 10) have increased the percentage of expenses in R&D activities. Austria is among the countries that shares the largest percentage of spending from the group of the countries in the study. From 1.8% of expenses in 2000, it has increased to 3.07% in 2016. Denmark is another country that spends quite large amount of the GDP in research and development activities. From 1981, it has almost tippled the amount from: 1.01% in 1981 to 2.97 after three decades, respectively, in 2011. Luxembourg, a small country with the largest GDP per capita (after Monaco, that is not included in the study since no data were available for this indicator), is the only country in this group that has decreased the share of expenses for R&D purposes, from 1.57% in 2000 to 1.24 in 2016.

Table 4.2 R&D spending as % of GDP by countries with the largest GDP per capita

No.	Country	1981	1991	2000	2011	2016	+ Increased investments / – Decreased investments
1	Luxembourg	/	/	1.576729	1.462771	1.243661	–
2	Switzerland	1.973506176	/	2.323444	/	/	+
3	Norway	1.154424637	1.593518	1.553457	/	/	+
4	Iceland	0.644401499	1.327399	2.732489	/	/	+
5	Ireland	0.648097348	0.893564	1.084969	1.55049	1.17681	+
6	Singapore	/	/	1.821545	2.145855	/	+
7	Denmark	1.013748355	1.583289	2.441446	2.970482	/	+
8	Sweden	2.057512405	3.913822	3.249186	/	/	+
9	Australia	/	/	1.57596	2.24574	/	+
10	Austria	/	/	1.89139	2.67843	3.07181	+

Source The World Bank (https://data.worldbank.org/)

4.3.3 Expenditures in R&D from the Balkan Countries

The main idea to have comparison among developed and economically powerful countries and Balkan countries is to analyze from different perspective and contribution in R&D that might lead to the increase of economic growth. Another reason of getting compared the countries is to have the comparison among regional countries of Kosovo.

Since many of the Balkan countries have passed through unstable political and economical circumstances, there is lack of data for those countries for the periods before 1981–2001. Especially for those countries that were under Ex-Yugoslavia government: Montenegro, Bosnia and Herzegovina, and Kosovo, while data for Greece were available only up to 2003. Therefore, the periods for analysis are from 2003 to 2015. Nevertheless, available data enable us to analyze whether the investments are comparable to the previous groups of countries and those data enable us to see the trend of investments, whether it tends to go upward or downward (Table 4.3).

It is noticeable that the investments are far behind the developed countries of the two groups analyzed above: first group—countries with the largest GDP, and second group—the largest GDP per capita. Even the GDP of those countries is much larger than majority of Balkan countries (exception are Croatia, Slovenia, and Bulgaria), the average expenses as percentage of GDP were between 1.5 and 2.5%, while in the Balkan countries, it is almost under 1% (only Slovenia with 2.19% in 2015). Also, percentage of the countries that are decreasing investments in R&D is higher: 3 out of 5 countries. For the countries that have no data available, we cannot conclude if they are continuing to invest or might have no investments for R&D, e.g., Greece, since it is undergoing through very unstable economic situation. The country that has invested more after Slovenia is Bulgaria and Croatia with 0.95% which marks

Table 4.3 R&D spending as % of GDP by Balkan countries

No.	Country	2003	2008	2015	+ Increased investments / – Decreased investments
1	Kosovo	/	0.23	0.15	–
2	Albania	/	0.15421	/	/
3	Serbia	0.52145	0.71082	0.8661	+
4	Macedonia	0.21041	0.22352	0.44329	+
5	Montenegro	0.79774	/	0.38019	–
6	Bosnia and Herzegovina	0.01854	0.01843	0.21897	+
7	Croatia	0.95216	0.88333	0.85439	–
8	Bulgaria	0.47379	0.42683	0.95657	+
9	Greece	0.96636	/	/	/
10	Slovenia	1.24528	1.625639	2.196553	+

Source The World Bank (https://data.worldbank.org/)

a significant increase from the last period in 2008, and Serbia has marked a slightly positive trend on investments, from 0.71% in 2008 to 0.86% in 2015.

Kosovo stands very weak at this point. Even the level of expenses was very low in 2008, only 0.23% of the GDP spent in R&D, it is decreased to 0.15% in 2015. Almost same for Albania for 2008 which was only 0.15% and is followed by Bosnia and Herzegovina with 0.21% in 2015, which actually was even worse in 2003 and 2008 with level of investments only 0.01%.

4.3.4 Policies for the Investments in R&D at the European Level

Benefits from the investments in research and development are of particular interest for economy. European trends on investing in R&D are becoming more and more ambitious. The initial strategy at the European level, known as Lisbon Strategy, where set goals, which among others are greater participation in R&D and ICT among and thus implement reforms to improve education system, stimulates knowledge and innovation toward a stronger industrial basis (Rodriguez et al., 2010). The three pillars that strategy has are: (a) dynamic, competitive, knowledge-based economy; (b) investments in human resources and social inclusion; (c) balanced use of natural resources and economic growth. In this context, at the European level, science and technology remain among the core fields for development. This strategic document aimed to serve as a booster for European Union countries—EU, for increasing the level of investment in research and development. Other reasons that are vital for economic growth are (EuropeanCommission, 2003):

- To help European companies innovate and stay competitive
- To create more and better jobs in Europe
- To keep improving the European way of life.

These initiatives for increasing R&D investments by European Union countries, were also based on good examples of in order of the USA 2.5% and Japan 3%. Low level of participation in R&D by the EU countries is mainly as a result of the lack of private investments. Therefore, Europe should be focused on improving conditions for investments in R&D by private sector (European Commission, 2010).

The same strategy was followed by the Barcelona European Council objective that was determined by 2010, to raise the R&D investments up to 3% of GDP. The initiatives were focused on achieving objectives of improving employment and growth in Europe in the long-run perspective. According to the forecast by that time, these investments by the threshold of 3% would bring in the future—after 2010, a GDP growth up to 0.5% per year and around 400,000 jobs each year. Unfortunately, it was not achieved, since the GDP growth of the countries was lower than the investments countries needed to invest to achieve the objective by 2010 (European Commission, 2003). Additionally, "more resources for R&D and innovation are a necessity but they are an insufficient means to achieve the goal of an Innovative Europe. A paradigm change is needed in which European values are preserved but in a new social structure. Productivity of R&D must be increased. The proportion of structural funds spent on research and innovation should be trebled" (Aho et al., 2006) since the R&D spending by the European Union countries barely reached 2% (Fig. 4.1).

Although, action plans were developed and revised periodically to achieve the goals, R&D investments marked stagnation. During the period of 2014–2016, they reported to be in average 2.03 % of the GDP. Therefore, the target of achieving 3% of GDP is still to be reached by 2020, (Eurostat, 2018) facilitated by private investments as well. Additionally, the commission proposes to also develop indicator that would not only reflects on R&D, but also on innovation intensity (European Commission, 2010) (Fig. 4.2).

This can be seen in the figure above, up to 2009, the investments by the European Union countries in average were 2.1% in 2009, while in 2000 investments were 1.86%. This means that even the agenda of EU was to increase more level of the investments by countries, the total average increase for ten years was quite low, and quite far behind the objectives, with only around 0.2%.

In the Europe 2020 strategy, key components are considered R&D and innovation, to prepare and strengthen the economy for the decade of 2010–2020. Targets of the EU are focused in five fields (Eurostat, 2018):

- Employment
- Research and development
- Climate change and energy

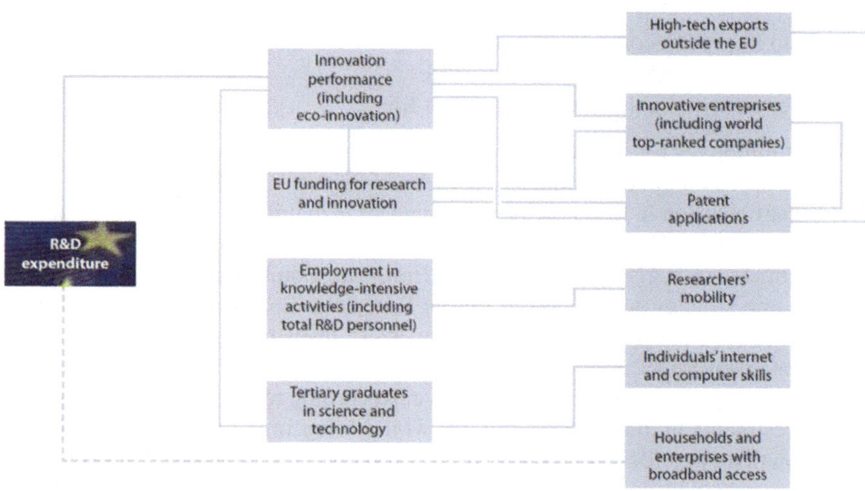

Fig. 4.1 Indicators and concepts on the R&D target. *Source* Eurostat statistics explained (https://ec.europa.eu/eurostat)

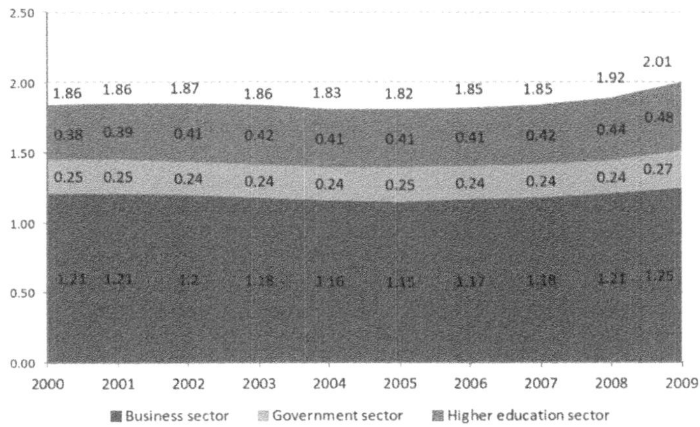

Fig. 4.2 Expenditure of R&D as percentage of GDP in EU 27, 2000–2009. *Source* The Europe 2020 Strategy, 2012 (Gros & Roth, 2012)

- Education
- Poverty and social exclusion.

According to the strategy "Innovative products and services not only contribute to the strategy's smart growth goal but also to its inclusiveness and sustainability objectives. Moreover, introducing new ideas to the market promotes industrial competitiveness, job creation, labour productivity and the efficient use of resources. R&D and innovation are also essential for finding solutions to societal challenges such as climate change, clean energy, security and active and healthy ageing" (European Commission, 2010). Detailed description on the objectives and initiatives of the strategy 2020 is given in the table (Table 4.4).

The projections for achieving goals of R&D spending by the European Union countries for 2020 are shown in the figure (Fig. 4.3):

Even the goals to increase expenditure were set in the early 2000, the R&D investments declined from 2000 up to 2007. From being 1.86% in 2000, they were slightly declined in the next following years to 2007 to 1.85%. A slight increase is seen in the last two years of measurement by reaching 1.92% in 2009 and 2.01% in 2010. Consequently, more investments are needed to be dedicated to R&D by each year, in order to reach the target of 3% in 2020 and stay competitive with other world developed countries.

Table 4.4 Europe 2020: an overview

– Headline targets		
Raise the employment rate of the population aged 20–64 from the current 69% to at least 75% Achieve the target of investing 3% of GDP in R&D in particular by improving the conditions for R&D investment by the private sector and develop a new indicator to track innovation Reduce greenhouse gas emissions by at least 20% compared to 1990 levels or by 30% if the conditions are right, increase the share of renewable energy in our final energy consumption to 20%, and achieve a 20% increase in energy efficiency Reduce the share of early school leavers to 10% from the current 15% and increase the share of the population aged 30–34 having completed tertiary education from 31% to at least 40% Reduce the number of Europeans living below national poverty lines by 25%, lifting 20 million people out of poverty		
– Smart growth	– Sustainable growth	– Inclusive growth
Innovation EU flagship initiative "Innovation Union" to improve framework conditions and access to finance for research and innovation so as to strengthen the innovation chain and boost levels of investment throughout the union	*Climate, energy and mobility* EU flagship initiative "Resource efficient Europe" to help decouple economic growth from the use of resources, by decarbonizing our economy, increasing the use of renewable sources, modernizing our transport sector, and promoting energy efficiency	*Employment and skills* EU flagship initiative "An agenda for new skills and jobs" to modernize labor markets by facilitating labor mobility and the development of skills throughout the life cycle with a view to increase labor participation and better match labor supply and demand
Education EU flagship initiative "Youth on the move" to enhance the performance of education systems and to reinforce the international attractiveness of Europe's higher education	*Competitiveness* EU flagship initiative "An industrial policy for the globalisation era" to improve the business environment, especially for SMEs, and to support the development of a strong and sustainable industrial base able to compete globally	*Fighting poverty* EU flagship initiative "European platform against poverty" to ensure social and territorial cohesion such that the benefits of growth and jobs are widely shared and people experiencing poverty and social exclusion are enabled to live in dignity and take an active part in society

Source EUROPE 2020. A strategy for smart, sustainable and inclusive growth, 2010 (https://eur-lex.europa.eu/legal-content/EN/ALL/?uri=CELEX:52010DC2020)

4.3.5 World Top R&D Investments Distribution

Investments on R&D can lead to wide network of the innovations such as: processes or design of the new products. These are protected by the IP rights Industrial Property (rights). "IP portfolios were identified using matching procedures linking data on the name of patent, trademark and design assignees to the names of the top corporate R&D investors and their subsidiaries. Top R&D investors in the ICT industries present relatively more concentrated IP portfolios in terms of both technologies (patents) and products (trademarks and designs)" (Daiko et al., 2017). The figure

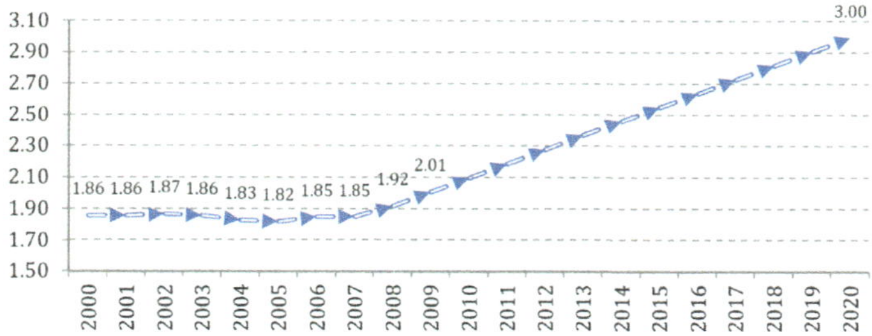

Fig. 4.3 Expenditure of R&D as percentage of GDP in EU 27, 2000–2009—projection to reach Europe 2020 target. *Source* The Europe 2020 strategy, 2012 (Gros & Roth, 2012)

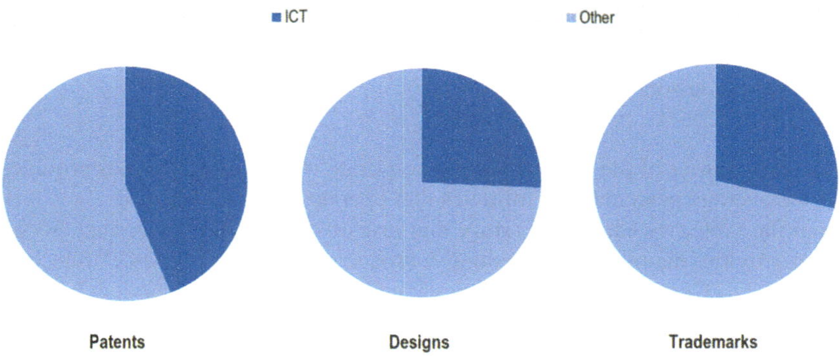

Fig. 4.4 ICT-related IP rights owned by world top R&D performers, 2012–2014. *Source* European commission's joint research centre & organisation for economic co-operation and development (OECD), "World corporate top R&D investors: industrial property strategies in the digital economy"

below shows percentage of the top R&D performers in the world in ICT-related patents, designers, and trademarks during the period 2012–2014 (Fig. 4.4).

R&D investors that operate in the ICT sectors share more than 25%, while the majority of the R&D investors (more than 70%) operate in the USA, Japan, Chinese, Chinese Taipei. Only the USA shares about 29% of world top investments in R&D, and 41% of them are related to ICT industries. While the same study conducted by (Daiko et al., 2017) with 2000 R&D companies, reported that in 2014, the companies were more spread geographically, where the majority of them about 60% (1119 companies) had the headquarters in the USA, the UK, and Germany, whereas in China and Chinese Taipei were 15% (297 companies). Around forth of total (571 companies out of 2000) operated in the ICT industry.

Anyhow, the other percentage of the R&D investments goes to the other sectors and industrial diversification that varies across economic sectors. The figure below shows details on the investments in ICT industries, and on the other sectors, as defined by ISIC rev.4, for the year of 2014. The figure gives an insight to the orientation of

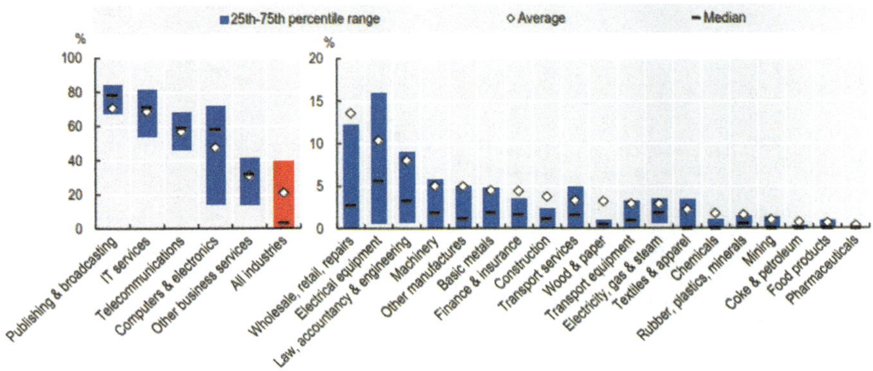

Fig. 4.5 ICT subsidiaries of the world's top R&D investors, by industry, ISIC rev. 4, 2014. Percentage of ICT subsidiaries. *Source* European commission's joint research centre & organisation for economic co-operation and development (OECD), "World corporate top R&D investors: industrial property strategies in the digital economy"

the ICT subsidiaries of the R&D investors corporate worldwide. Statistics provided shows the average percentage of affiliates that operates in ICT (Fig. 4.5).

From the table, we can easily notify that investments in ICT are spread among different activities mainly on: "Publishing & broadcasting", "IT services", "Telecommunications", "Computers & electronics", and "other business services". Noteworthy is the fact that the 197 "Pharmaceutical" companies in the sample do not seem to rely on affiliates operating in ICT. Likewise, three of the most populated sectors, i.e., "machinery" (153 companies), "chemicals" (132 companies), and "transport equipment" (146 companies), also had very low percentage values of affiliates operating in ICT, respectively, 5%, 3%, and 2%. ICT industries also show the greatest dispersion across companies, as illustrated by the 25th–75th percentile range (Daiko et al., 2017).

4.4 Conclusions

In the chapter, we explored the phases from R&D investments to technological innovation and diffusion. It is quite easy to define each phase, but in practical way, it is very difficult to distinguish among phases and realize the whole process. Beside the money and human recourses, the time distance is very long. Depending on the type of invention, sometimes it takes some years, and sometimes it can even take hundred years to employ in practice the invention. Due to the complexity of the new products/processes, it is depended on countries investments whether they can close the circle of all phases of the research and development. Usually, it is dependent on the level of investments in each phase. Meaning, less investments or not, the probability to finalize research is low, and the opposite, more investments, country can succeed to even diffuse the invention at the broader level.

When looking from the point of probability at getting any result from the investments, chances are lower at the beginning—early phases of research cycle, respectively, in the phase of fundamental research. This also tells that the uncertainty at the beginning is higher, and the uncertainty at the other phases becomes lower and lower going through diffusion. Expressed in percentage, 78% of the scientific research have probability to be transformed into result, while in particular, fundamental research chances are only 5–10% to be transformed into final result. Applicative researches are more likely to become implemented 85–90%, while the investments in development researches have 95–97% probability to be transformed into the expected result (Dedaj, 1998).

Although the importance of R&D plays crucial role in the technological progress, thus it can be seen as facilitator of the economic development. On the other hand, not all of countries have possibility to develop such researches (as it requires huge financial support and professional human capital), then it is leaded by the financially developed countries such as: the USA.

Therefore, the importance of the researches is not only in economic aspect but also in the technologic aspect. Based on the role and importance, the innovations can be divided in two groups: (a) primary and (b) secondary.

(a) Primary inventions: offers new technological solutions, when in most cases are as result of fundamental research. In this case, totally new technologies appear. These technologies bring new manufacturing activities or services.
(b) Secondary inventions: represents sophistication of tools and existing processes. In these cases, there are improvements on the primary inventions, but without changing main characteristics and functions.

Based on what was discussed and analyzed, we can conclude that technological progress is dependent on the investments. Thus, the strategies of the developed countries, such as European Union countries, are seeking to improve the ration of the expenditure in the R&D.

Even the strategy for reaching the level of 3% by the EU countries by 2010 was not made possible; in the analysis of the individual countries, we saw the trends of increase were evident at almost all EU countries. Hence, most countries are in general working toward achievement of the 2020 target, to reach 3% of the total expenditure of GDP. Although it seems to be quite difficult, as only few EU countries have already reached the level of 3% of investments. Countries such as: Austria, Germany, and France stand quite well and have come close to the limit—the lowest threshold.

Regarding to the Balkan countries, as most countries are developing countries, the level is quite low; thus, possibilities on a technological progress and economic growth are low. Slovenia, as member of European Union, stands very well when it comes to the R&D investments, while the rest in 2015 have invested less than 1% of the GDP, including Croatia that is also member of the European Union, as well as Greece. Much progress is needed and initiatives also by the Balkan countries to keep the trend world developments, which in fact they can beardly reach the threshold of 3% even in the next decade.

Countries that are referring point for the R&D have their headquarters in the USA—as the biggest investor in the world; in geographic context, they are equally spread in the two other continents. In Europe, they are located mostly in: Germany and the UK, and in Asia: China and Japan.

As regard to the dispersion of the R&D investments in economic sectors, the recent study conducted with top R&D worldwide performers with 2000 companies shows a close relationship of the investments in ICT sectors. It can be noticed from the Industrial Property (IP) rights that are relatively high in sector of ICT, designs, and less on the trademarks in comparison with other activities. Consequently, R&D investments seem to be headed on the digitalization and technological development worldwide.

References

Aho, E., Cornu, J., Georghiou, L., & Subirá, A. (2006). *Creating an innovative Europe*. European Commission.

Archibugi, D., & Coco, A. (2004). A new indicator of technological capabilities for developed and developing countries (ArCo). *World Development, 32*, 629–654. Elsevier.

Audretsch, D. B., Litan, R. E., & Strom, R. J. (2009). Industrial dynamics, entrepreneurship and innovation. In R. E. David & B. Audretsch (Eds.), *Massachusetts 01060*. Edward Elgar Publishing Limited.

Bender, G., & co. (2008). Innovation in low-tech firms and industries. In: H. H.-K. Jacobson (Ed.), *Massachusetts*. Edward Elgar Publishing Limited.

Bilbao-Osorio, B., & Rodriguez-Pose, A. (Fall 2004). Economic, from R&D to innovation and growth in the EU. *Growth and Change, 35*(4), 434–455.

Bruijn, T. D., & Norberg-Bohm, V. (2005). *Industrial transformation; environmental policy innovation in the United States and Europe*. The MIT Press–Massachusetts Institute of Technology.

Cantner, U., & Malerba, F. (2006). *Innovation, industrial dynamics and structural transformation: Schumpeterian legacies*. Springer.

Daiko, T., Dernis, H., Dosso, M., Gkotsis, P., Squicciarini, M., & Vezzani, A. (2017). *World corporate top R&D investors: Industrial property strategies in the digital economy*. European commission's joint research centre; Organisation for Economic Co-operation and Development (OECD).

Dedaj, B. (1998). *Roli i shkences dhe teknologjise ne zhvillimin ekonomik - naliza e disa aspekteve te pergjithshme*. Fakulteti Ekonomik, Universiteti i Prishtines.

Devetakovic, S. (1989). *Tehnicki progres u maloj zemlji*.

European Commission. (2003). *Investing in Europan reseach*. European Union.

European Commission. (2010). *Europe 2020. A strategy for smart, sustainable and inclusive growth*. EuropeanCommission.

Eurostat. (2018). *Eurostat*. Retrieved September 2018, from Europe 2020 indicators—R&D and innovation: https://ec.europa.eu/eurostat/statistics-explained/index.php/Europe_2020_indicators_-_R%26D_and_innovation#General_overview

Giacchero, A., Donnini, N., & Martin, F. (2007). Innovation and competitiveness in local SMEs: Characteristics of firms, entrepreneurs, environment and their interrelationships. In U. Cantner & F. Marelba (Eds.), *Innovation and the evolution of industries* (pp. 174–189). Springer.

Godin, B. (2006). The linear model of innovation: The historical construction of an analytical framework. *Science, Technology, and Human Values, 31*(6), 639–667.

Gregorio, D. D., & Shane, S. (2003). Why do some universities generate more start-ups than others? *Elsevier, Research Policy, 32*(2003), 209–227.

Gros, D., & Roth, F. (2012). *The Europe 2020 strategy. Can it maintains the EU's competitiveness in the world?* Centre for European Policy Studies (CEPS).

Ignat, V. (2017). *Digitalization and the global technology trends.* IOP Publishing.

Lin, J. Y. (2012). *New structural economics: A framework for rethinking development.* Washington DC 20433. The World Bank.

Loschel, A. (2002). Technological change in economic models of environmental policy: A survey. *Ecological Economics, 43,* 105–126. Elsevier.

Maclaurin, W. R. (1953, February). The sequence from invention to innovation and its relation to economic growth. *The Quarterly Journal of Economics, 67*(1), 97–111.

Mesaric, M. (1971). *Suvremena znanstveno tehnicka revolucija.* Ekonomski institut.

Mowery, D. C., & Rodenberg, N. (1998). *Paths of innovation. Technological change in 20th-century America.* Cambridge University Press. https://www.amazon.com/Paths-Innovation-Technological-20th-Century-America-dp-0521641195/dp/0521641195/ref=mt_other?_encoding=UTF8&me=&qid=

Mueller, W. F. (1962). The rate and direction of inventive activity: Economic (pp. 323–358.

Nelson, R. R. (1993). *National innovation systems. A comparative analysis.* Oxford University Press.

Norman, D. A., & Verganti, R. (2013). Incremental and radical innovation: Design research versus technology and meaning change. *30*(1). Winter 2014.

Pasinetti, L. L. (1993). *Structural economic dynamics: A theory of the economic consequences of human learning.* Cambridge University Press.

Rodriguez, R., Warmerdam, J., & Triomphe, C. (2010). *The Lisbon strategy 2000–2010. An analysis and evaluation of the methods used and results achieved.* European Parliament, Policy Department Economic and Scientific Policy;. European Parliament.

Rodríguez-Pose, A., & Crescenzi, R. (2010, October 7). Research and development, spillovers, innovation systems, and the genesis of regional growth in Europe. *Regional Studies, 42*(1), 51–67.

Schumpeter, J. A. (1947, November). The creative reponse in Economic Histrory. *The Journal of Economic History, 7*(2), 149–159.

Schumpeter, J. A. (1943). *Capitalism, socialism and democracy.* Taylor & Francis.

Souoza, P. D. (2010). *Innovation in industrial research.* CSIRO Publishing.

UNIDO. (2016). *The role of technology and innovation in inclusive and sustainable industrial development.* United Nations Industrial Development Organization.

United Nations. (2008). *International standard industrial classification of all economic activities, revision 4.* United Nations Publication.

Chapter 5
Analysis of CSR—Corporate Social Responsibility—Initiatives by CPSEs—Central Public Sector Enterprises of India

Jagbir Singh Kadyan, M. Abu Naser, and Leena Jenefa

Abstract In India, Central Public Sector Enterprises—CPSEs—are the government-owned corporations in which majority (51% or more) of the paid share capital is either held by central government or by any state government or partly by the central government and partly by one or more state governments. According to the Government of India, Public Enterprise Survey Report, 2018–19, there were total of 178 profit-making CPSEs making a net profit of ₹1,74,587 crore. The top ten ranking profit-making CPSEs among them made a net profit of ₹1,07,947 crore, contributing 61.83% to the total profit earned by the CPSEs during 2018–19. CSR—Corporate Social Responsibility—in India is mandatory in nature. According to Section 135 of Companies Act, 2013, all profit-making corporates, including Central Public Sector Enterprises (CPSEs), exceeding threshold limits prescribed in the Act regarding net worth of ₹500 crore (₹5000 million), or turnover of ₹1000 crore (₹10,000 million) or net profit of ₹5 crore (₹50 million), in pursuance of its CSR policy are mandated to spend at least 2% of the average net profit (profit before tax) of immediate three preceding year ₹ towards CSR activities as per items listed in Schedule VII of the Companies Act 2013. Such companies are required to constitute a CSR committee, formulate a CSR policy and undertake CSR initiatives in accordance with Schedule VII of the Act and are also required to mandatorily report it in their annual reports in the prescribed format. The mandatory CSR reporting facilitates the company to demonstrate their commitment towards the society and also act as a communication tool to engage with different stakeholders, including shareholders, regulators, communities, customers and society at large. Central Public Sector Enterprises (CPSEs) in India are regularly undertaking various social initiatives and fulfil their corporate social responsibility towards the

J. S. Kadyan (✉)
Department of Commerce, Swami Shraddhanand College, University of Delhi, New Delhi, India
e-mail: dr.jskadyan@gmail.com; Prof.jskadyan@gmail.com

M. A. Naser
London Metropolitan University, London, UK

L. Jenefa
Vel Tech University, Chennai, India

M. A. Naser (ed.), *New Approaches to CSR, Sustainability and Accountability, Volume IV*, Accounting, Finance, Sustainability, Governance & Fraud: Theory and Application, https://doi.org/10.1007/978-981-16-9499-8_5

society and the nation. This research paper analyses the corporate social initiatives undertaken by the top ten profit-making CPSEs of India as per their ranking in the Public Enterprise Survey report, 2018–19. The duration of this research work is for the period of 05 year starting from the financial year 2014–15 to 2017–19.

Keywords Corporate social responsibility · Central public sector enterprises · Indian companies act 2013 · NITI Aayog · Society · MCA-ministry of corporate affairs · Sustainable development & society

Abbreviations

CPSE Central Public Sector Enterprises
MCA Ministry of Corporate Affairs

5.1 Introduction

The Concept of CSR—Corporate Social Responsibility—is not new to India. For ages, entrepreneurs and businesspersons have tried to help the society through some or other means in its overall development. Various charitable and philanthropic activities were done for the society. The word used for was not CSR, but the spirit was that of CSR.[1] The ancient Indian wisdom of the social distribution of wealth is mainly guided by the "Vedas",[2] which attest that there will be an appropriate conveyance of wealth from the well off to poor people and censure individuals who appreciate wealth without sharing it with others within the society.

With the complete overhaul of erstwhile Company Act 1956 and the introduction of Company Act 2013, India has witnessed remarkable changes in the society and the corporates operating within the society. Corporates have seriously started assessing and re-assessing their social responsibilities, not only towards their immediate stakeholders but also towards the society in which they have been functioning ever since they were incorporated.

Similarly, academicians, sociologist and researchers have also become active in their respective spears of work encircling around the society and CSR—Corporate Social Responsibility. The concept of CSR was introduced by Companies Act, 2013 with the insertion of Section 135, which deals with the provisions related to CSR. Ministry of Corporate Affairs (MCA), one of the corporate regulatory bodies of India, vide its notification dated 27 February 2014, notify the provisions of CSR shall be with effective from 1 April 2014.

---

[1] D. Crowther, S. Seifi (eds.), The Palgrave Handbook of Corporate Social Responsibility, https://doi.org/10.1007/978-3-030-22438-7_86-1.

[2] "Vedas", The Ancient holy Indian Scriptures. https://www.ancient.eu/The_Vedas.

With the introduction of Section 135 of Companies Act 2013, CSR is mandatory in India. Accordingly, any company having a turnover of ₹1000 crores (₹10,000 millions) or net worth of ₹500 crores (₹5000 millions) or net profit of ₹5 crores (₹50 millions) during the preceding 3 years are mandatorily required to spent 2% of their net profits towards CSR.[3] The CSR spending is to be done in accordance with the activities enumerated in Schedule VII of the Companies Act 2013.[4]

Accordingly, various companies falling within the ambit of Section 135 have begun to identify various CSR initiatives that can be undertaken by them and accordingly have started making budgetary allocations for the same. Committees and task forces have been set up by the companies to identify their thrust areas and formulate CSR policies in accordance to the Schedule VII of the Company Act 2013, which has specifically specified twelve crore areas/components/items on which the CSR initiatives of the specified companies are required to formulate.

CPSEs—Central Public Sector Enterprises—are the government-owned corporations in which majority (51% or more) of the paid share capital is either held by central government or by any state government or partly by the central government and partly by one or more state governments. Over a period of time, CPSEs have been making a significant contribution towards the development of the nation.

5.2 Corporate Social Responsibility in India

The enactment of CSR in India with the introduction of Section 135 of the Company Act 2013 was a major milestone for India. In the absence of Section 135, corporates were making social contributions towards the society as per their choice and preferences. Most of them were guided by different philosophies, beliefs, experiences, culture, traditions and social systems. Therefore, with the introduction of mandatory CSR, it paved a way for guided and directional approach towards planning and execution of CSR projects in India. The following were the major milestones towards enactment of CSR in India (Table 5.1).

As discussed earlier, any company in India having a "net worth" of ₹500 crore or more, or "turnover" of ₹1000 crore or more, or "net profit" of ₹5 crore or more, during the immediately three preceding financial year are required to spent 2% of the annual average net profits towards CSR in accordance with Schedule VII of the Company Act 2013. All such companies are required to constitute a CSR committee of the Board,[5] comprising of three or more directors, out of which at least one director shall be an independent director. However, in case where the appointment of independent director is not applicable, then such company shall have only 2 or more directors in such committee.

[3] Section 135 (i) Company Act 2013: https://www.mca.gov.in/Ministry/pdf/CompaniesAct2013.pdf.

[4] Section 135 Company Act 2013; https://www.mca.gov.in/Ministry/pdf/CompaniesAct2013.pdf.

[5] Section 135 (3) of Company Act 2013.

Table 5.1 Major milestones towards enactment of CSR in India

Dates	Major events
Dec. 2009	Proposal for making voluntary provisions for CSR under the company bill
8 Sept. 2010	Parliament standing committees' recommendation for mandatory CSR
18 Dec. 2012	Company bill passed in Lok Sabha
8 Aug. 2013	Company bill passed in Rajya Sabha
13 Aug. 2013	SEBI requirement for disclosure of CSR spending by the largest 100 Cos
29 Aug. 2013	Company bill received assent of the President of India
27 Feb. 2014	The companies (CSR Policy) rules, 2014 Notification by ministry of corporate affairs, MCA
1 April 2014	Applicability of Section 135 containing provisions regarding CSR

Source Compilation by the researcher

The CSR committee shall be responsible towards formulating and recommending the CSR policy to the board, indicating the various CSR activities to be undertaken by the company in accordance with the Schedule VII of the Company Act 2013. The CSR Committee shall also regularly review and monitor the CSR policies and the activities undertaken by the company.

Similarly, the Companies (CSR Policy) Rules, 2014,[6] notified on 27 Feb. 2014 and which were effective from 1 April 2014 are mandatory in nature and are required to be complied with by all the companies, including the CPSEs.

5.3 Schedule VII of Company Act 2013, (Section 135)[7]

Schedule VII of the Companies Act 2013 forms the basis of CSR spending in India. The entire CSR programme of the company had to be based on it. Any activities undertaken by the companies not falling within Schedule VII does not qualify as CSR activity in India. The Company Law 2013 clearly states the activities which may be included by companies in their Corporate Social Responsibility Programme.

The following are the CSR activities according to Schedule VII of the Co. Act 2013. Activities relating to:

S-i. [Eradicating hunger, poverty, and malnutrition],[8] ["promoting health care including preventive health care"][9] and sanitation [including contribution to

[6] The Companies (Corporate Social Responsibility Policy) Amendment Rules, 2020. https://www.mca.gov.in/Ministry/pdf/csr_26082020.pdf.

[7] Companies Act 2013, Section 135.

[8] Substituted by Notification Dated 27 February 2014.

[9] Substituted by Notification Dated 31 March 2014. For the words "promoting preventive health care" read "promoting health care including preventive health care".

the Swachh Bharat Kosh set up by the central government for the promotion of sanitation] and making available safe drinking water.

S-ii. Promoting education, including special education and employment, enhancing vocation skills, especially among children, women, elderly and the differently abled and livelihood enhancement projects.

S-iii. Promoting gender equality, empowering women, setting up homes and hostels for women and orphans; setting up old-age homes, day care centres and such other facilities for senior citizens and measures for reducing inequalities faced by socially and economically backward groups.

S-iv. Ensuring environmental sustainability, ecological balance, protection of flora and fauna, animal welfare, agroforestry, conservation of natural resources and maintaining quality of soil, air and water [including contribution to the Clean Ganga Fund set up by the central government for rejuvenation of river Ganga].[10]

S-v. Protection of national heritage, art and culture including restoration of buildings and sites of historical importance and works of art; setting up public libraries; promotion and development of traditional art and handicrafts;

S-vi. Measures for the benefit of armed forces veterans, war widows and their dependents;

S-vii. Training to promote rural sports, nationally recognised sports, Paralympics sports and Olympic sports.

S-viii. Contribution to the prime minister's national relief fund or any other fund set up by the central govt. for socio-economic development and relief and welfare of the schedule caste, tribes, other backward classes, minorities and women;

S-ix. Contributions or funds provided to technology incubators located within academic institutions which are approved by the central govt.

S-x. Rural development projects.

S-xi. [Slum area development. Explanation: For the purposes of this item, the term 'slum area' shall mean any area declared as such by the central government or any state government or any other competent authority under any law for the time being in force].[11]

S-xii. [Disaster management, including relief, rehabilitation and reconstruction activities].[12]

These are the twelve sets of CSR activities mentioned in the Schedule VII of the Companies Act 2013. Any company falling within the purview of Section 135 of the Companies Act 2013 are mandatorily required to undertake only those activities which are mentioned in the Schedule VII of the Companies Act 2013 while planning their CSR activities.

Companies are required to form a CSR committee for the purpose of planning and implementing their CSR programme for the year. The CSR committee of the

[10] Inserted by Notification Dated 24 October 2014.

[11] Inserted by Notification Dated 7 August 2014.

[12] Inserted by Notification dated 30th may 2019.

Company shall select the activities listed within Schedule VII only in order to formulate their CSR programmes and the allocation of the budgets for the same. The CSR committee shall be submitting the CSR proposal and the budgeted allocation to the BOD for their approval and sanctions. The Board of Directors upon receiving the CSR committees' proposal may approve and sanction it for its execution and implementation for the year.

5.4 About CPSEs—Central Public Sector Enterprises

In India, CPSEs—Central Public Sector Enterprises—are the government-owned corporations in which majority (51% or more) of the paid share capital is either held by central government or by any state government or partly by the central government and partly by one or more state governments. According to the Government of India, Public Enterprise Survey Report-2018–19, there were a total of 178 profit-making CPSEs, making a net profit of ₹1,74,587 crore. The top ten ranking profit-making CPSEs among them made a net profit of ₹1,07,947 crore, contributing 61.83 %to the total profit earned by the CPSEs during 2018–19.

According to the Press Information Bureau, the Government of India, there were total 348 CPSEs as on 31 March 2019, out of which 249 were operational. Remaining 86 CPSEs were under construction and 13 CPSEs were under closure or liquidation. The highlights of the performance of CPSEs, during 2018–19 are as follows[13]:

- Total paid-up capital in all CPSEs as on 31.3.2019 stood at ₹2,75,697 crore as compared to ₹ crore as on 31.3.2018, showing a growth of 8.55%.
- Total financial investment in all CPSEs stood at ₹16,40,628 crore as on 31.3.2019 compared to ₹14,31,008 crore as on 31.3.2018, recording a growth of 14.65%.
- Capital employed in all CPSEs stood at ₹26,33,956 crore on 31.3.2019 compared to ₹23,57,913 crore as on 31.3.2018, showing a growth of 11.71%.
- Total gross revenue from operation of all CPSEs during 2018–19 stood at ₹25,43,370 crore compared to ₹21,54,774 crore in the previous year showing a growth of 18.03%.
- Total income of all CPSEs during 2018–19 stood at ₹24,40,748 crore compared to ₹20,32,001 crore in 2017–18, showing a growth of 20.12%.
- Profit of 178 profit-making CPSEs stood at ₹1,74,587 crore during 2018–19 compared to ₹1,55,931 crore in 2017–18, showing a growth in profit by 11.96%.
- Loss of 70 loss making CPSEs stood at ₹31,635 crore in 2018–19 compared to ₹32,180 crore in 2017–18, showing decrease in losses by 1.69%.
- Overall net profit of operating CPSEs during 2018–19 stood at ₹1,42,951 crore as compared to ₹1,23,751 crore during 2017–18, showing a growth in overall profit of 15.52%.

[13] https://pib.gov.in/PressReleseDetailm.aspx?PRID=1602623.

- Reserves and surplus of CPSEs stood at ₹9,93,328 crore as on 31.03.2019 as compared to ₹9,26,906 crore as on 31.03.2018, showing an increase by 7.17%.
- Net worth of all CPSEs went from ₹11,15,552 crore as on 31.03.2018 to ₹12,08,758 crore as on 31.03.2019, showing an increase of 8.36%.
- Contribution of CPSEs to Central Exchequer by way of excise duty, customs duty, GST, corporate tax, interest on central government loans, dividend and other duties and taxes stood at ₹3,68,803 crore in 2018–19 as against ₹3,52,361 crore in 2017–18, showing an increase of 4.67%.
- Foreign exchange earnings of 79 CPSEs through exports of goods and services stood at ₹1,43,377 crore in 2018–19 against ₹98,714 crore in 2017–18, showing an increase of 45.24%.
- Foreign exchange expenditure of 144 CPSEs on imports and royalty, know-how, consultancy, interest and other expenditure stood at ₹6,64,914 crore in 2018–19 against ₹5,22,256 crore in 2017–18, showing an increase of 27.32%.

During the year 2018–19, out of the total 348 CPSEs, 178 were profit-making CPSEs, making a profit of ₹1,74,587 crore. Similarly, the top 10 profit-making CPSEs selected for the study had made a profit amounting to ₹1,07,947 crore, contributing 61.83 %[14] of the total profit earned by all the profit-making CPSEs for the year.

The list of top ten profit-making CPSEs selected for this research work is mentioned below.

1. ONGC—Oil and Natural Gas Ltd.
2. IOC—Indian Oil Corporation Ltd.
3. NTPC—Natural Thermal Power Corporation Ltd.
4. CIL—Coal India Ltd.
5. PGCL—Power Grid Corporation of India Ltd.
6. BPCL—Bharat Petroleum Corporation Ltd.
7. PFCL—Power Finance Corporation Ltd.
8. MCL—Mahanadi Coalfields Ltd.
9. HPCL—Hindustan Petroleum Corporation Ltd.
10. GAIL—Gas Authority of India Ltd.

5.5 Research Methodology

Data collection for this research work is primarily done from the annual reports, CSR reports of the selected companies. Similarly, reports of the Department of Public Enterprises, Ministry of Corporate Affair, National Stock Exchange and other regulatory bodies have been utilised. MS-Excel has been used to present, process and analyse the data to arrive at appropriate findings and conclusions.

The CSR expenditures data of each company has been tabulated year wise, as per Schedule VII of the Company Act 2013, which contains XII items of expenditure. Every company has to structure their CSR activities and programmes their

[14] Public Enterprise Survey Report, GOI, 2018–19.

allocations based on it. The following process has been adopted for the data analysis, interpretations and conclusions for this research study.

(i) **CSR Budgets, CSR Spending and Over/Under-CSR spending by the CPSEs of India**:

Under this heading, we are highlighting and analysing the CSR initiatives undertaken by the CPSEs of India coming under the ambit of Section 135 of the Company Act 2013. Consolidated CSR budget, CSR spending and over/underspending of CSR of all the CPSEs of India towards the twelve components of the Schedule VII of the Companies Act 2013 has been highlighted and analysed. Year-wise analysis has been done for a period of five years, starting from 2014–15 onwards till 2018–19.

(ii) **Allocation towards CSR Budget by the selected 10 CPSE's**:

Under this heading, we are highlighting and analysing the allocation of CSR budget by the selected 10 CPSEs towards the twelve components of the Schedule VII of the Company Act 2013. Consolidated as well as individual comparative analysis has been done among the selected 10 CPSEs with regard to the allocation of their CSR budgets. Year-wise analysis has been done for a period of five years, starting from 2014–15 onwards till 2018–19.

(iii) **CSR Spending by the selected 10 CPSEs**:

Under this heading, we are highlighting and analysing the CSR spending done by the selected 10 CPSEs towards the twelve components of the Schedule VII of the Company Act 2013. Consolidated as well as individual comparative analysis has been done among the selected 10 CPSEs with regard to their CSR spending. Year-wise analysis has been done for a period of five years, starting from the year 2014–15 onwards till 2018–19.

(iv) **Over/Underspending on CSR by the selected 10 CSPEs**:

Under this heading, we are highlighting and analysing the over- or underspending done by the selected 10 CPSEs towards the twelve components of the Schedule VII of the Company Act 2013. Consolidated as well as individual comparative analysis has been done among the selected 10 CPSEs with regard to over- or underspending on CSR. Year-wise analysis has been done for a period of five years, starting from 2014–15 onwards till 2018–19.

5.6 Data Analysis and Interpretations

Data collection is primarily done from the CPSE's annual reports, CSR reports, Public Enterprise Survey reports, Ministry of Corporate Affairs, National Stock Exchange and other statutory regulatory bodies for the purpose of this research work.

At the beginnings of this research work, data analysis and interpretation of the CSR initiatives undertaken by the CPSEs of India towards the 12 components of Schedule

VII of the Company Act 2013 has been done, in order to ascertain the yearly changes in the pattern of their CSR spending. The current year (2018–19) CSR initiatives undertaken by the CPSEs of India has been compared with the previous year (2017–18) only, since the paucity of the requisite data for the same. The variations observed are highlighted accordingly in the form of yearly change expressed in the form of percentage.

Subsequently, consolidated CSR initiatives undertaken by the selected 10 CPSEs in accordance with the Schedule VII of the Company Act 2013 has been interpreted and analysed. Similarly, inter-CPSEs comparison of the allocation of the CSR budgets, CSR initiatives undertaken and the over-/underspending towards the CSR initiatives undertaken by the selected 10 CPSEs towards the 12 components of the Schedule VII of Company Act 2013 has been interpreted and analysed accordingly.

5.6.1 CSR Spending by the CPSEs of India

CSR spending by the CPSEs in India according to the Schedule VII of Company Act 2013 is available only from the year 2017 to 18. The details are as under. During the year 2017–18, a total of 153 CPSE were coming within the ambit of Section 135 of the Co. Act 2013. The total amount of 2% of PBT, Profit Before Taxes, of these CPSEs was amounting to ₹3693.47 crore. However, an amount of ₹3442.42 crore only was found to be actually spent on their CSR activities. Thus, there was underspending on CSR activities to the tone of ₹251.05 crore. The highest amount of CSR spent according to the different items of Schedule VII was towards item S-I, amounting to ₹1119.34, while the lowest amount of ₹0.36 crore was spent towards S-ix. Similarly, no amount was spent for item no. S-xii of the Schedule VII of the Company Act 2013 (Table 5.2).

During the year 2018–19, a total 158 CPSEs were coming within the ambit of Section 135 of the Co. Act 2013. The total amount of 2% of PBT, Profit Before Taxes, of these CPSEs was amounting to ₹3620.47 crore, and the allocation of funds towards CSR initiatives was amounting to ₹5686.75 (which includes the carried forward amount of the previous years). However, an amount of ₹3873.32 crore only was found to be actually spent on their CSR activities. Therefore, there was underspending on CSR activities to the tone of ₹1813.43 crore. The highest amount of CSR spent according to different items of Schedule VII was towards item S-i amounting to ₹1333.83, while the lowest amount of ₹0.77 crore was towards S-ix and no amount was spent towards item no. S-xii of the Schedule VII of the Company Act 2013.

Out of the twelve items of Schedule VII, there has been an increasing trend towards CSR spending on nine out of twelve items of the Schedule VII of the Company Act 2013, and similarly, there has been a decreasing trend towards CSE spending on three out of twelve items of Schedule VII of the company act 2103. However, from the total CSR spending of all CPSEs put together, there is an increasing trend of CSR spending of 12.52%, amounting to ₹430.90 crore for the year 2018–19.

Table 5.2 CSR spending CPSEs of India as per schedule VII of company act 2013

Schedule VII (components)	CSR spending done on	2017–18 Amount ₹ in crore	2018–19 Amount ₹ in crore	Yearly change in %
S-i	Eradicating hunger and poverty, health care and sanitation	1119.34	1333.83	**19.16**
S-ii	Education and skill development	1112.65	1259.41	**13.19**
S-iii	Empowerment of women and other economically backward sections	66.14	67.29	**1.74**
S-iv	Environmental sustainability	393.47	374.55	**−4.81**
S-v	Art and culture	244.32	64.6	**−73.56**
S-vi	Armed forces welfare	8.5	15.15	**78.24**
S-vii	Sports	64.98	57.54	**−11.45**
S-viii	Contribution to funds set up by central government	50.92	122.96	**141.48**
S-ix	Contribution to technology incubators	0.36	0.77	**113.89**
S-x	Rural development	374.68	563.26	**50.33**
S-xi	Slum area development	7.06	13.96	**97.73**
S-xii	Disaster management, including relief, rehabilitation and reconstruction activities			
	Total	**3442.42**	**3873.32**	**12.52**

Note Detail break-up of CSR spending for the period from 2014–15 to 2016–17 is not available, hence not considered
Source https://dpe.gov.in/publicenterprises-survey-2018-19

5.6.2 Consolidated CSR Budget, Spending and Over/Underspending by the Selected 10 CPSEs

Under this heading of the research work, we have attempted to interpret and analyse the consolidated CSR budget, the consolidated CSR spending and the over/underspending done by the selected 10 CPSEs in accordance with Schedule VII of the Company Act 2013.

While the five years consolidated CSR budget of all the selected 10 CPSEs towards the Schedule VII of the Company Act 2013 was amounting to ₹16,019.88 crore (100%), the consolidated actual CSR spending done by all the selected 10 CPSEs towards the Schedule VII of the Company Act 2013 was amounting to ₹8,799.85 crore (55%). Therefore, it can be observed that there is consolidated underspending

Table 5.3 Consolidated CSR budget, spending and over/underspending by the selected 10 CPSEs

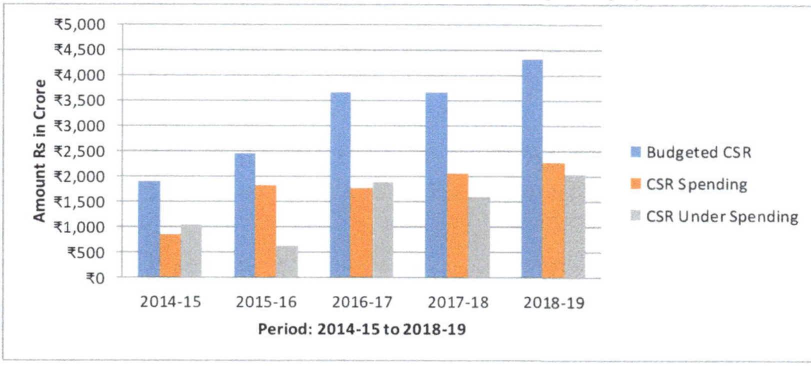

Source Compilation done by the researcher

on the CSR initiatives, amounting to ₹7,220.30 crore (45%) for the five-year period of the study starting from the year 2014–15 to 2018–19.

During the year 2014–15, according to Table 5.3 above, the budgeted CSR allocation by the selected 10 CPSEs was amounting to ₹1906.07 crore (100%). The actual CSR spending done by all the selected 10 CPSEs was only to the extent of 44.61% of their budgeted CSR allocation, amounting to ₹850.23 crore. Thus, the selected 10 CPSEs had underspent their CSR budget to the extent of 55.39% of their budgeted CSR allocation amounting to ₹1055.84 crore.

During the year 2015–16, according to Table 5.3 above, the budgeted CSR allocation by the selected 10 CPSEs was amounting to ₹2463.30 crore (100%). The actual CSR spending done by the selected 10 CPSEs was to the extent of 77.40% of their budgeted CSR allocation, amounting to ₹1833.09 crore. Thus, the selected 10 CPSEs had underspent their CSR budget to the extent of 22.59% of their budgeted CSR allocation, amounting to ₹630.21 crore.

During the year 2016–17, according to Table 5.3 above, the budgeted CSR allocation by the selected 10 CPSEs was amounting to ₹3664.39 crore (100%). The actual CSR spending done by the selected 10 CPSEs was only to the extent of 50.13% of their budgeted CSR allocation, amounting to ₹1769.33 crore. Thus, the selected 10 CPSEs had underspent their CSR budget to the extent of 49.87% of their budgeted CSR allocation amounting to ₹1895.06 crore.

During the year 2017–18, according to Table 5.3 above, the budgeted CSR allocation by the selected 10 CPSEs was amounting to ₹3661.83 crore (100%). The actual CSR spending done by the selected 10 CPSEs was only to the extent of 57% of their budgeted CSR allocation, amounting to ₹2076.16 crore. Thus, the selected 10 CPSEs had underspent their CSR budget to the extent of 43% of their budgeted CSR allocation amounting to ₹1585.67 crore.

During the year 2018–19, according to Table 5.3 above, the budgeted CSR allocation by the selected 10 CPSEs was amounting to ₹4324.29 crore (100%). The actual CSR spending done by the selected 10 CPSEs was only to the extent of 52.52% of their budgeted CSR allocation, amounting to ₹2271.04 crore. Thus, the selected 10 CPSEs had underspent their CSR budget to the extent of 47.48% of their budgeted CSR allocation, amounting to ₹2053.25 crore.

5.6.3 Consolidated CSR Budget as Per Schedule VII of the Company Act 2013 by the Selected 10 CPSEs

Under this heading, the interpretation and analysis of the selected 10 CPSE's consolidated CSR budget towards the Schedule VII of the Company Act 2013, which comprises of xii items/components, has been done. CPSEs are mandatorily required to undertake only those CSR initiatives which are mentioned the Schedule VII of the Company Act 2013. Thus, the allocation of consolidated CSR budget towards the different items/components of Schedule VII has been identified individually as well as collectively and analysed accordingly for the five-year period starting from the year 2014–15 to 2018–19.

Schedule VII of the Company Act 2013 contains only twelve components/items. All organisations are therefore allowed to choose and spend only towards these twelve components under their CSR initiatives. However, it is observed that the selected 10 CPSEs has additionally chosen two more components/items, viz. "Others" and "Administrative Overheads", under their CSR initiatives. Thus, the number of components/items of Schedule VII of the Company Act 2013 sums up to fourteen. Therefore, for this research work, it is assumed that the Schedule VII of the Company Act contains 14 components/items instead of 12 and these additional two components/items are also shown separately as "Others" and "Administrative Overheads".

According to Table 5.4 **above, during the year 2014–15**, the consolidated budgetary allocation towards CSR by all the selected 10 CPSEs was found to be ₹1906.07 crore, which was budgeted to be spent on the 14 different items of the Schedule VII of the Company Act 2013. While the average consolidated budgetary CSR allocation was found to be ₹173.28 crore, the highest collective budgetary allocation of ₹736.60 crore was found to be one towards S-i, while the lowest collective budgetary allocation of ₹1.49 crore was found to be done towards S-xi item of Schedule VII of the Company Act 2013. Out of the total 14 items from the Schedule VII, only 11 items were preferred while three items, namely S-vi, S-ix and S-xii, were not found to be preferred.

During the year 2015–16, as per Table 5.3 above, the Consolidated Budgetary Allocation towards CSR by all the selected 10 CPSEs was found to be ₹2463.94 crore, which was budgeted to be spent on the 11 different items of the Schedule VII of the Company Act 2013. While the average consolidated budgetary CSR allocation was found to be ₹223.94 crore, the highest collective budgetary allocation of ₹1214.29 crore was found to be one towards S-i, while the lowest collective budgetary allocation of ₹0.50 crore was found to be done towards S-vi item of Schedule VII of the

Table 5.4 Consolidated CSR budget of the selected 10 CPSEs as per schedule VII of company act 2013

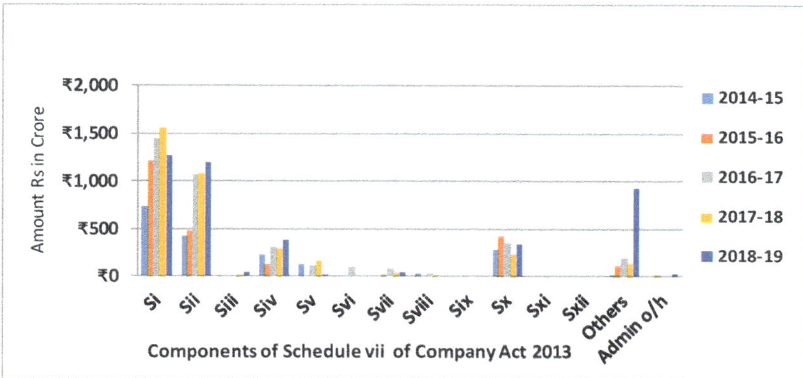

Source Compilation done by the researcher

Company Act 2013. Out of the total 14 items from the Schedule VII, eleven items were preferred while three items, namely S-ix, S-xi and S-xii, were not found to be preferred.

During the year 2016–17, as per Table 5.3 above, the consolidated budgetary allocation towards CSR by all the selected 10 CPSEs was found to be ₹3664.39 crore, which was budgeted to be spent on the 12 different items of the Schedule VII of the Company Act 2013. While the average consolidated budgetary CSR allocation was found to be ₹305.37 crore, the highest collective budgetary allocation of ₹1446.26 crore was found to be one towards S-i, while the lowest collective budgetary allocation of ₹0.75 crore was found to be done towards S-xii item of Schedule VII of the Company Act 2013. Out of the total 14 items from the Schedule VII, twelve items were preferred, while two items, namely S-ix and S-xi, were not found to be preferred.

During the year 2017–18, as per Table 5.3 above, the consolidated budgetary allocation towards CSR by all the selected 10 CPSEs was found to be ₹3661.83 crore, which was budgeted to be spent on the 12 different items of the Schedule VII of the Company Act 2013. While the average consolidated budgetary CSR allocation was found to be ₹305.15 crore, the highest collective budgetary allocation of ₹1565.04 crore was found to be one towards S-i, while the lowest collective budgetary allocation of ₹1.75 crore was found to be done towards S-xii item of Schedule VII of the Company Act 2013. Out of the total 14 items from the Schedule VII, twelve items were preferred while two items, namely S-ix and S-xi, were not found to be preferred.

During the year 2018–19, as per Table 5.3 above, the consolidated budgetary allocation towards CSR by all the selected 10 CPSEs was found to be ₹4323.29 crore, which was budgeted to be spent on the 13 different items of the Schedule VII of the Company Act 2013. While the average consolidated budgetary CSR allocation was found to be ₹332.64 crore, the highest collective budgetary allocation of ₹1271.50 crore was found to be one towards S-i, while the lowest collective budgetary allocation of ₹0.75 crore was found to be done towards S-vi item of Schedule VII of the Company Act 2013. Out of the total 14 items from the Schedule VII, thirteen items were preferred while only one item, namely S-ix, was not found to be preferred.

5.6.4 Consolidated CSR Spending as Per Schedule VII of the Company Act 2013 by the Selected 10 CPSEs

Under this heading, the interpretation and analysis of the selected 10 CPSE's consolidated CSR spending towards the Schedule VII of the Company Act 2013 has been done. As mentioned earlier, CPSEs are mandatorily required to undertake only those CSR initiatives which mention the Schedule VII of the Company Act 2013. The consolidated CSR spending done towards the different items/components of Schedule VII has been identified individually as well as collectively and analysed accordingly for the five-year period starting from the year 2014–15 to 2018–19.

Schedule VII of the Company Act 2013 contains only twelve components/items. All organisations are therefore allowed to choose and spend only towards these twelve components under their CSR initiatives. However, it is observed that the selected 10 CPSEs have additional chosen two more components/items, viz. "Others" and "Administrative Overheads", under their CSR initiatives. Thus, the number of components/items of Schedule VII of the Company Act 2013 sums up to fourteen. Therefore, for this research work, it is assumed that the Schedule VII of the Company Act contains 14 components/items instead of 12 and these additional two components/items are also shown separately as "Others" and "Administrative Overheads".

During the year 2014–15, as per Table 5.4 above, the actual consolidated CSR spending towards CSR by all the selected 10 CPSEs was found to be ₹850.23 crore, which was actually spent on the eleven different items of the Schedule VII of the Company Act 2013. While the average actual consolidated CSR spending was found to be ₹77.29 crore, the highest consolidated CSR spending was found to be ₹337.88 crore which was found to be spent towards S-iv, while the lowest consolidated CSR spending of ₹1.35 crore was found to be spending towards item S-xi of Schedule VII of the Company Act 2013. Out of the total 14 items from the Schedule VII, eleven items were preferred, while three items, namely S-vi, S-ix and S-xii, were not found to be preferred for CSR spending.

During the year 2015–16, as per Table 5.4 above, the actual consolidated CSR spending towards CSR by all the selected 10 CPSEs was found to be ₹1833.09 crore, which was actually spent on the twelve different items of the Schedule VII of the Company Act 2013. While the average actual consolidated CSR spending was found to be ₹152.76 crore, the highest consolidated CSR spending was found to be ₹984.21 crore which was found to be spent towards S-iv, while the lowest consolidated CSR spending of ₹1.35 crore was found to be spending towards item S-xi of Schedule VII of the Company Act 2013. Out of the total 14 items from the Schedule VII, twelve items were preferred, while two items, namely S-ix and S-xi, were not found to be preferred for CSR spending.

During the year 2016–17, as per Table 5.4 above, the actual consolidated CSR spending towards CSR by all the selected 10 CPSEs was found to be ₹1769.33 crore, which was actually spent on the twelve different items of the Schedule VII of the Company Act 2013. While the average actual consolidated CSR spending was found to be ₹147.44 crore, the highest consolidated CSR spending was found to be ₹579.16 crore which was found to be spent towards S-ii, while the lowest consolidated CSR spending of ₹0.10 crore was found to be spending towards item S-vi of Schedule VII of the Company Act 2013. Out of the total 14 items from the Schedule VII, twelve items were preferred while two items, namely S-ix and S-xi, were not found to be preferred for CSR spending.

During the year 2017–18, as per Table 5.4 above, the actual consolidated CSR spending towards CSR by all the selected 10 CPSEs was found to be ₹2076.16 crore, which was actually spent on the twelve different items of the Schedule VII of the Company Act 2013. While the average actual consolidated CSR spending was found to be ₹173.01 crore, the highest consolidated CSR spending was found to be ₹746.15 crore which was found to be spent towards S-i, while the lowest consolidated CSR spending of ₹1.75 crore was found to be spending towards item S-xii of Schedule VII of the Company Act 2013. Out of the total 14 items from the Schedule VII, twelve items were preferred, while two items, namely S-ix and S-xi, were not found to be preferred for CSR spending.

During the year 2018–19, as per Table 5.4 above, the actual consolidated CSR spending towards CSR by all the selected 10 CPSEs was found to be ₹2271.04 crore, which was actually spent on the twelve different items of the Schedule VII of the Company Act 2013. While the average actual consolidated CSR spending was found to be ₹189.25 crore, the highest consolidated CSR spending was found to be ₹611.82 crore which was found to be spent towards "others", while the lowest consolidated CSR spending of ₹0.75 crore was found to be spending towards item S-vi of Schedule VII of the Company Act 2013. Out of the total 14 items from the Schedule VII, twelve items were preferred while two items, namely S-viii and S-ix, were not found to be preferred for CSR spending.

5.6.5 Consolidated Over/Underspending of CSR as Per Components of Schedule VII of Company Act 2013 by the Selected 10 CPSEs

Under this heading, over/underspending of CSR as per the Schedule VII of the Company Act 2013 by the selected 10 CPSEs has been done. The consolidated over/underspending of CSR spending done towards the different items/components of Schedule VII has been identified individually as well as collectively and analysed accordingly for the five-year period starting from the year 2014–15 to 2018–19.

Schedule VII of the Company Act 2013 contains only twelve components/items. All organisations are therefore allowed to choose and spend only towards these twelve components under their CSR initiatives. However, it is observed that the selected 10 CPSEs have additional chosen two more components/items, viz. "Others" and "Administrative Overheads", under their CSR initiatives. Thus, the number of components/items of Schedule VII of the Company Act 2013 sums up to fourteen. Therefore, for this research work, it is assumed that the Schedule VII of the Company Act contains 14 components/items instead of 12 and these additional two components/items are also shown separately as "Others" and "Administrative Overheads".

According to Table 5.6 above, **during the year 2014–15**, it has been observed that the consolidated budgetary allocation towards CSR spending by the selected CPSEs was found to be done towards eleven items only out of the total fourteen items of the Schedule VII of the Company Act 2013. Overspending was observed towards the three items, namely S-iv, vii and "Administrative Overheads". Similarly, underspending was observed on the eight items, namely S-i, S-ii, S-iii, S-v, S-viii, S-x, S-xi and "Others".

During the Year 2015–16, as per Table 5.6 above, it has been observed that the consolidated budgetary allocation towards CSR spending by the selected CPSEs were found to be done towards eleven items only out of the total fourteen items of the Schedule VII of the Company Act 2013. Overspending was observed towards the one item only, namely S-xi. Similarly, underspending was observed towards ten items, namely S-i, S-ii, S-iii, S-iv, S-v, S-vii, S-viii, S-x, S-xi, "Others" and "Administrative Overheads".

During the Year 2016–17, as per Table 5.6 above, it has been observed that the consolidated budgetary allocation towards CSR spending by the selected CPSEs were found to be done towards eleven items only out of the total fourteen items of the Schedule VII of the Company Act 2013. Overspending was observed towards the one item only, namely "Administrative Overheads". Similarly, underspending was observed towards ten items, namely S-i, S-ii, S-iii, S-iv, S-v, S-vi, S-vii, S-viii, S-x and "Others".

During the Year 2017–18, as per Table 5.5 above, it has been observed that the consolidated budgetary allocation towards CSR spending by the selected CPSEs were found to be done towards ten items only out of the total fourteen items of the Schedule VII of the Company Act 2013. Overspending was observed towards the one item only, namely S-v. Similarly, underspending was observed towards nine items, namely S-i, S-ii, S-iii, S-iv, S-vi, S-vii, S-viii, S-x and "Others".

During the Year 2018–19, as per Table 5.6 above, it has been observed that the consolidated budgetary allocation towards CSR spending by the selected CPSEs were found to be done towards ten items only out of the total fourteen items of the Schedule VII of the Company Act 2013. Over Spending was observed towards the one item only, namely "Administrative Overheads". Similarly, Under Spending was observed towards nine items, namely S-i, S-ii, S-iii, S-iv, S-v, S-vii, S-x, S-xi and "Others".

Table 5.5 Consolidated CSR Spending by selected 10 CPSE's as per Components of Schedule vii of Company Act 2013

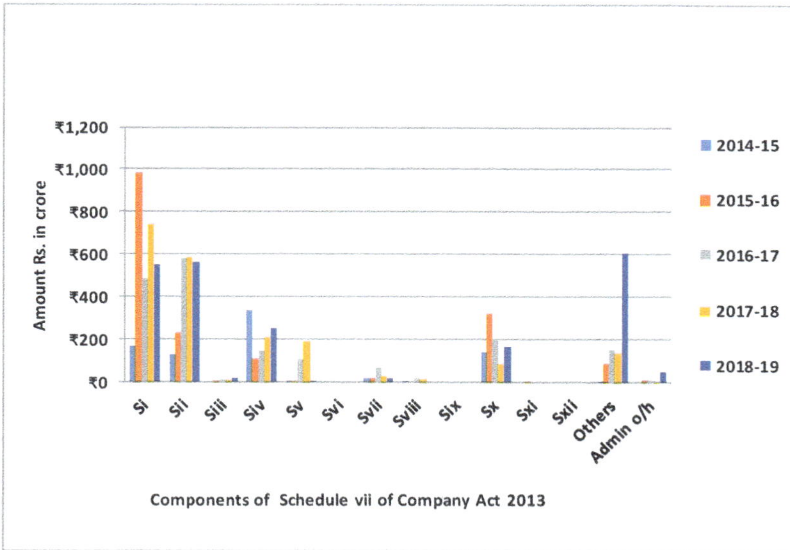

Source Compilation done by the researcher

Table 5.6 Consolidated over/underspending of CSR by the selected 10 CPSEs as per components of schedule VII of company act 2013

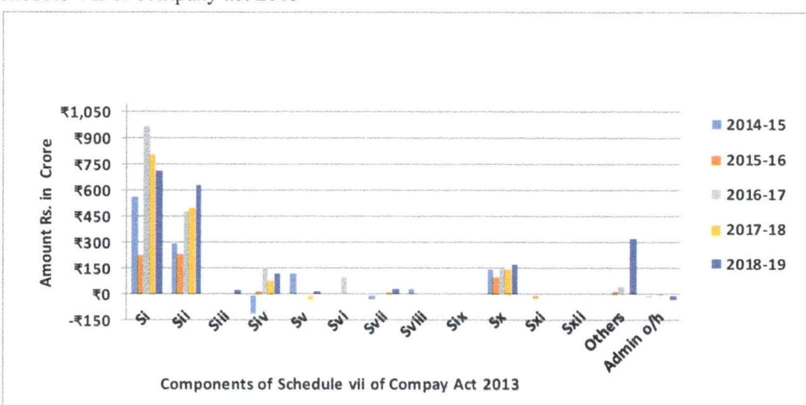

Source Compilation done by the researcher

5.6.6 *Inter-CPSEs Analysis of Allocation of CSR Budget Towards Sch. Vii*

Under this heading, we have attempted to analyse the allocation of the CSR budget towards the various component of Schedule VII of the Company Act 2013 by the selected 10 CPSEs. Year-wise analysis has been done for all the selected 10 CPSEs collectively as well as individually, for the period of five years starting from the year 2014–15 to 2018–19.

During the year 2014–15, as per Table 5.7 above, the **collective budgetary allocation** towards CSR by all the selected 10 CPSEs was found to be ₹1906.07 crore, which was budgeted to be spent on the 14 different items of the Schedule VII of the Company Act 2013. While the average collective budgetary CSR allocation towards CSR was found to be ₹173.28 crore, the highest collective budgetary allocation of ₹736 crore was found to be done towards S-i, while the lowest collective budgetary allocation of ₹1.49 crore was found to be done towards S-xi item of Schedule VII of the Company Act 2013.

While the average budgetary CSR allocation done by the **selected 10 CPSEs** towards Schedule VII items of CSR was ₹211.79 crore, ONGC Ltd. had the highest budgetary CSR allocation of ₹660.60 crore budgeted towards five items of the Schedule VII, and GAIL Ltd. had the lowest budgetary CSR allocation of ₹2.19 crore budgeted towards only three items of the Schedule VII of the Company Act 2013.

The average budgetary CSR allocation towards different **items of Schedule VII** of the Company Act 2013 was PGC Ltd. made the highest number of budgetary CSR allocations towards eight items of Schedule VII and GAIL made budgetary CSR

Table 5.7 Allocation of CSR budget by the selected 10 CPSEs towards Schedule VII of the company act 2013 for the year 2014–15

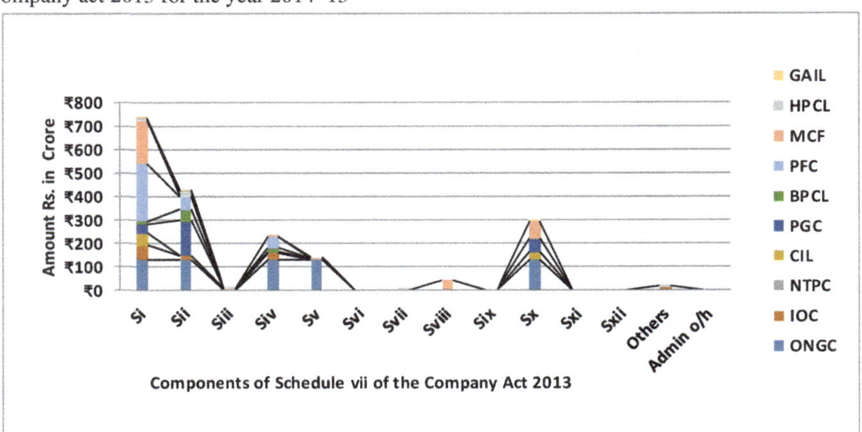

Source Compilation done by the researcher

allocation towards only three items of the Schedule VII of the Company Act 2013. The relevant data of NTPC Ltd. was not available for the year 2014–15; hence, it is not considered in the data analysis for the said year.

The most preferred item for the budgetary CSR allocation from the Schedule VII was S-i and S-ii, nine out of the ten CPSEs preferred them, and the least preferred item was S-xi, which was preferred by only one CPSE. Similarly, the most neglected item of Schedule VII was S-vi, S-ix and S-xii, which none of the selected CPSEs preferred.

During the year 2015–16, as per Table 5.8 above, the **collective budgetary allocation** towards CSR by all the selected 10 CPSEs was found to be ₹2449.48 crore, which was budgeted to be spent on the 14 different items of the Schedule VII of the Company Act 2013. While the average collective budgetary CSR allocation towards CSR was found to be ₹222.68 crore, the highest collective budgetary allocation of ₹1228.46 crore was found to be done towards S-i, while the lowest collective budgetary allocation of ₹0.50 crore was found to be done towards S-vii item of Schedule VII of the Company Act 2013.

While the average budgetary allocation done by the **selected 10 CPSEs** towards Schedule VII items of CSR was ₹246.12 crore, ONGC Ltd. had the highest budgetary CSR allocation of ₹525.71 crore budgeted towards nine items of the Schedule VII and HPCL had the lowest budgetary CSR allocation of ₹71.76 crore budgeted towards six items of the Schedule VII of the Company Act 2013.

The average budgetary CSR allocation towards different **items of Schedule VII** of the Company Act 2013 was five. ONGC Ltd. had selected a maximum of nine

Table 5.8 Allocation of CSR budget by the selected 10 CPSEs towards schedule VII of the company act 2013 for the year 2015–16

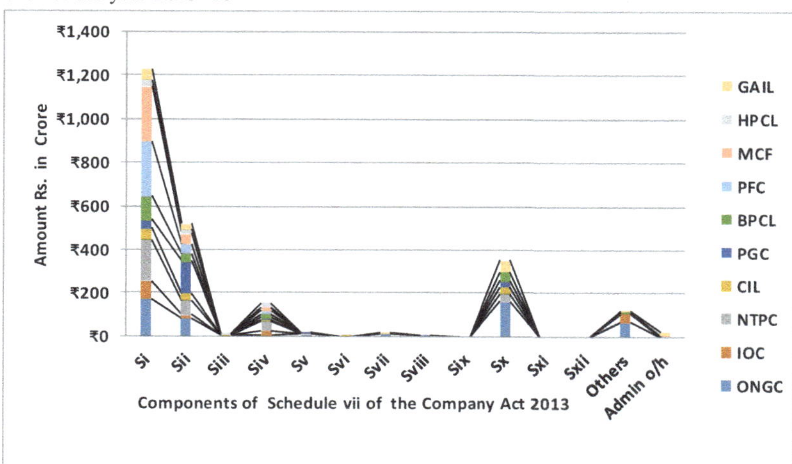

Source Compilation done by the researcher

items from Schedule VII and three CPSEs, namely IOC, PFC and MCF Ltd., selected the minimum of five items from the Schedule VII of the Company Act 2013.

The most preferred item for the budgetary CSR allocation from the Schedule VII were S-i and S-ii, all the ten CPSEs preferred them and the least preferred items were S-vi and S-viii, which was preferred by two CPSEs, namely CIL and ONGC. Similarly, the totally neglected items of the Schedule VII were S-ix, S-xi and S-xii, which none of the selected CPSEs preferred.

During the year 2016–17, as per Table 5.9 above, the **collective budgetary allocation** towards CSR by all the selected 10 CPSEs was found to be ₹2218.13 crore, which was budgeted to be spent on the 14 different items of the Schedule VII of the Company Act 2013. While the average collective budgetary CSR allocation towards CSR was found to be ₹305.37 crore, the highest collective budgetary allocation of ₹1446.26 crore was found to be done towards S-i, while the lowest collective budgetary allocation of ₹0.75 crore was found to be done towards S-xii item of Schedule VII of the Company Act 2013.

While the average budgetary allocation done by the **selected 10 CPSEs** towards Schedule VII items of CSR was ₹547.78 crore, ONGC Ltd. had the highest budgetary CSR allocation of ₹858.04 crore budgeted towards eight items of the Schedule VII and HPCL had the lowest budgetary CSR allocation of ₹108.11 crore budgeted towards six items of the Schedule VII of the Company Act 2013.

The average budgetary CSR allocation towards different **items of Schedule VII** of the Company Act 2013 was six. ONGC Ltd. had selected a maximum of eight

Table 5.9 Allocation of CSR budget by the selected 10 CPSEs towards schedule VII of the company act 2013 for the year 2016–17

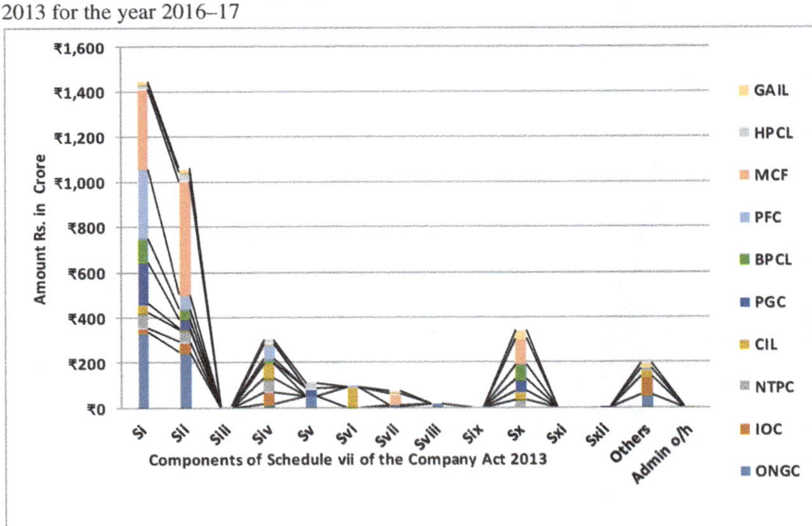

Source Compilation done by the researcher

items from Schedule VII and BPCL selected the minimum of four items from the Schedule VII of the Company Act 2013.

The most preferred item for the budgetary CSR allocation from the Schedule VII were S-i and S-ii, all the ten CPSEs preferred them, and the least preferred items were S-viii and S-xii, which were preferred by BPCL. Similarly, the totally neglected items of the Schedule VII were S-ix and S-xi, which none of the selected CPSEs preferred.

During the year 2017–18, as per Table 5.10 above, the **collective budgetary allocation** towards CSR by all the selected 10 CPSEs was found to be ₹3661.83 crore, which was budgeted to be spent on the 14 different items of the Schedule VII of the Company Act 2013. While the average collective budgetary CSR allocation towards CSR was found to be ₹305.15 crore, the highest collective budgetary allocation of ₹1565.04 crore was found to be done towards S-i, while the lowest collective budgetary allocation of ₹1.75 crore was found to be done towards S-xii item of Schedule VII of the Company Act 2013.

While the average budgetary allocation done by the **selected 10 CPSEs** towards Schedule VII items of CSR was ₹366.35 crore, ONGC Ltd. had the highest budgetary CSR allocation of ₹772.41 crore budgeted towards nine items of the Schedule VII, and CIL had the lowest budgetary CSR allocation of ₹64.55 crore budgeted towards seven items of the Schedule VII of the Company Act 2013.

The average budgetary CSR allocation towards different **items of Schedule VII** of the Company Act 2013 was seven. PGC Ltd. had selected a maximum of ten items from Schedule VII, and PFC selected the minimum of four items from the Schedule VII of the Company Act 2013.

The most preferred items for the budgetary CSR allocation from the Schedule VII were S-i and S-ii, all the ten CPSEs preferred them, and the least preferred item was

Table 5.10 Allocation of CSR budget by the selected 10 CPSEs towards schedule VII of the company act 2013 for the year 2017–18

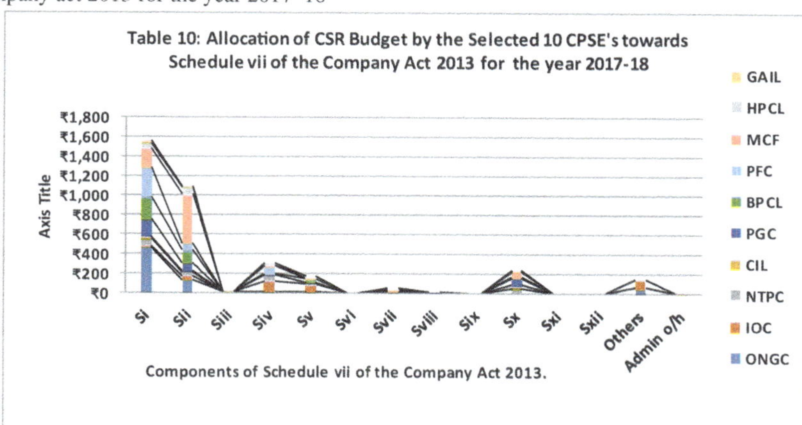

Source Compilation done by the researcher

S-xii, which was preferred only by PGC. Similarly, the totally neglected items of the Schedule VII were S-ix and S-xi, which none of the selected CPSEs preferred.

During the year 2018–19, as per Table 5.11 above, the **collective budgetary allocation** towards CSR by all the selected 10 CPSEs was found to be ₹4324.29 crore, which was budgeted to be spent on the 14 different items of the Schedule VII of the Company Act 2013. While the average collective budgetary CSR allocation towards CSR was found to be ₹360.36 crore, the highest collective budgetary allocation of ₹1271.50 crore was found to be done towards S-i, while the lowest collective budgetary allocation of ₹0.75 crore was found to be done towards S-vi item of Schedule VII of the Company Act 2013.

While the average budgetary allocation done by the **selected 10 CPSEs** towards Schedule VII items of CSR was ₹432.43 crore, ONGC Ltd. had the highest budgetary CSR allocation of ₹1271.50 crore budgeted towards seven items of the Schedule VII, and CIL had the lowest budgetary CSR allocation of ₹0.75 crore budgeted towards seven items of the Schedule VI of the Company Act 2013.

The average budgetary CSR allocation towards different **items of Schedule VII** of the Company Act 2013 was seven. PGC Ltd. had selected a maximum of nine items from Schedule VII and almost 50% of the CPSEs selected the minimum of six items from the Schedule VII of the Company Act 2013.

The most preferred item for the budgetary CSR allocation from the Schedule VII was S-i, all the ten CPSEs preferred them and the least preferred item was S-vi and S-xi, which was preferred only by PGC. Similarly, the totally neglected items of the Schedule VII were S-viii and S-xi, which none of the selected CPSEs preferred.

Table 5.11 Allocation of CSR budget by the selected 10 CPSEs towards schedule VII of the company act 2013 for the year 2018–19

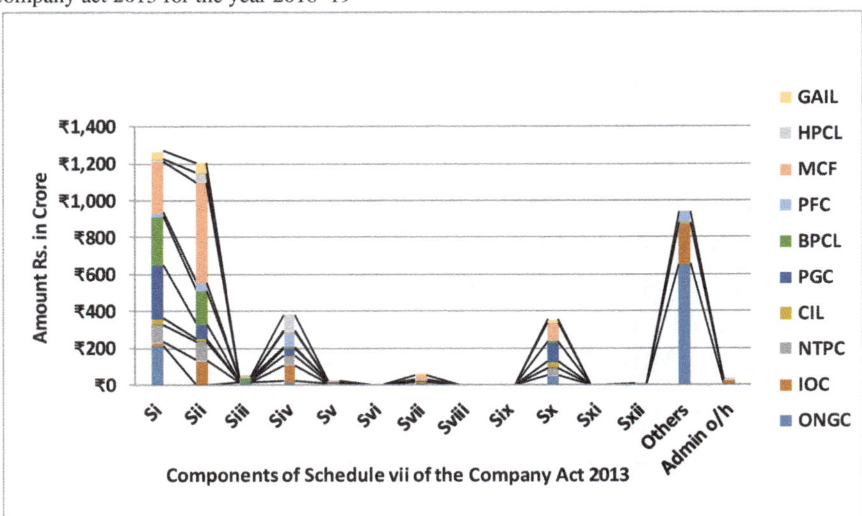

Source Compilation done by the researcher

5.6.7 *Inter-CPSEs Analysis of CSR Spending Towards Schedule VII of the Company Act 2013*

Under this heading, we have attempted to analyse the CSR spending done towards the various component of Schedule VII of the Company Act 2013 by the selected 10 CPSEs. Year-wise analysis has been done for all the selected 10 CPSEs collectively as well as individually, for the period of five years starting from the year 2014–15 to 2018–19.

During the year 2014–15, as per Table 5.12 above, the **Collective CSR Spending** done by the selected 10 CPSEs was found to be ₹850.23 crore, which was actually found to be spent on the eleven different items of the Schedule VII of the Company Act 2013. While the average actual CSR spending was found to be ₹77.29 crore, the highest CSR spending ₹337.88 crore done towards S-iv, while the lowest CSR spending of ₹1.35 crore was done towards item S-xi of Schedule VII of the Company Act 2013.

It is observed that, out of the total 14 items of the Schedule VII, the collective CSR spending was found to be done on 11 items, namely S-i, S-ii, S-iii, S-iv, S-v, S-vii, S-viii, S-x, S-xi and Others and Administrative Overheads. Similarly, no CSR spending was done on three items; namely, S-vi, S-ix and S-xii were not found to be done.

The average number of CSR activities undertaken for **CSR spending by the CPSEs** from the available fourteen items of Schedule VII of the Company Act 2013 was 05. The highest number of CSR activities undertaken was 09, led by PGC, and the least number of CSR activities undertaken was 03, led by GAIL. Since the details

Table 5.12 CSR spending by the selected 10 CPSEs towards schedule VII of the company act 2013 for the year 2014–15

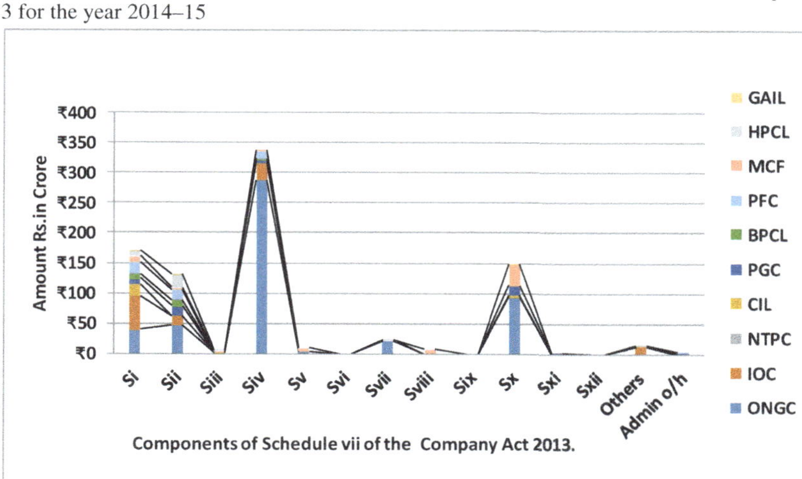

Source Compilation done by the researcher

about CSR spending done by NTPC was not available, it has not been considered in the above analysis.

The average number of CPSE's **CSR spending as per Schedule VII** of the company act 2013 was 04. The most preferred item of spending by the CPSEs from the Schedule VII was S-i, which was preferred by all the nine CPSEs, led by ONGC and followed by IOC Ltd and other. The least preferred item CPSEs from the Schedule VII were S-viii and S-xi, led by MCF and followed by PGC. Similarly, the Schedule VII items which were not considered by any of the CPSEs for their CSR spending for the year were S-vi, S-ix and S-xii.

During the year 2015–16, as per Table 5.13 above, the **collective CSR spending** done by the selected 10 CPSEs was found to be ₹1833.09 crore, which was actually found to be spent on the 12 different items of the Schedule VII of the Company Act 2013. While the average actual CSR spending was found to be ₹152.76 crore, the highest CSR spending ₹984.21 crore done towards S-i, while the lowest CSR spending of ₹0.50 crore was done towards item S-vi of Schedule VII of the Company Act 2013.

It is observed that, out of the total 14 items of the Schedule VII, the collective CSR spending was found to be done on 12 items, namely S-i, S-ii, S-iii, S-iv, S-v, S-vi, S-vii, S-viii, S-x, S-xi and Others & Administrative Overheads. Similarly, no CSR spending was done on two items, namely S-ix and S-xii, which were not found to be done.

The average number of CSR activities undertaken for **CSR spending by the CPSEs** from the available fourteen items of Schedule VII of the Company Act 2013

Table 5.13 CSR spending by the selected 10 CPSEs towards schedule VII of the company act 2013 for the year 2015–16

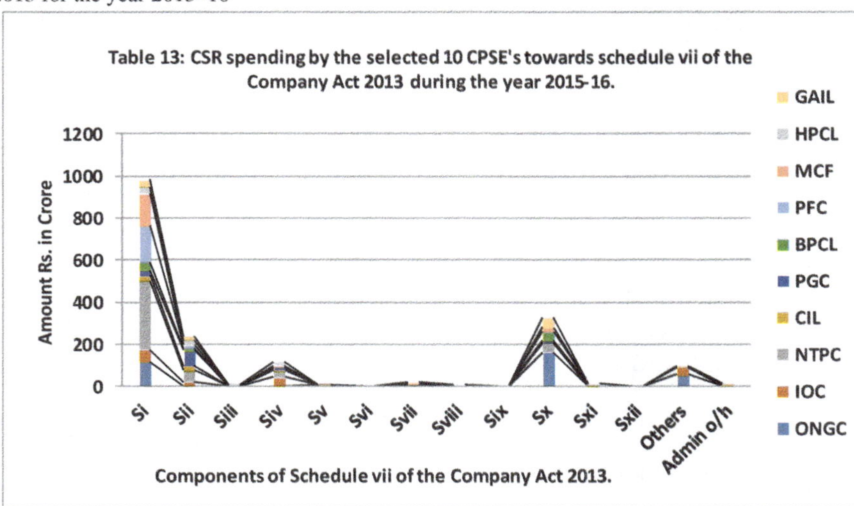

Source Compilation done by the researcher

was 07. The highest numbers of CSR activities undertaken was 09, led by CIL, and the least numbers of CSR activities undertaken was 05, led by PFC.

The average number of **CPSEs spending as per Schedule VII** of the Company Act 2013 was 05. The most preferred item of spending by the CPSEs from the Schedule VII was S-i, which was preferred by all the 10 CPSEs, led by NTPC. The least preferred item CPSEs from the Schedule VII were S-vi, led by CIL. Similarly, the items which were not considered by any of the CPSEs for their CSR spending for the year were S-v, S-viii and S-xii.

During the year 2016–17, as per Table 5.14 above, the **Collective CSR Spending** done by the selected 10 CPSEs was found to be ₹1769.33 crore, which was actually found to be spent on the 12 different items of the Schedule VII of the Company Act 2013. While the average actual CSR spending was found to be ₹147.44 crore, the highest CSR spending ₹579.16 crore was done towards S-ii, while the lowest CSR spending of ₹0.75 crore was done towards item S-xii of Schedule VII of the Company Act 2013.

It is observed that out of the total 14 items of the Schedule VII, the collective CSR spending was found to be done on 12 items, namely S-i, S-ii, S-iii, S-iv, S-v, S-vi, S-vii, S-viii, S-x, S-xii and Others & Administrative Overheads. Similarly, no CSR spending was done on two items, namely S-ix and S-xi, which were not found to be done.

The average number of CSR activities undertaken for **CSR spending by the CPSEs** from the available fourteen items of Schedule VII of the company act 2013 was 07. The highest numbers of CSR activities undertaken was 09, led by PGC and the least numbers of CSR activities undertaken was 05, by BPCL.

Table 5.14 CSR spending by the selected 10 CPSEs towards schedule VII of the company act 2013 for the year 2016–17

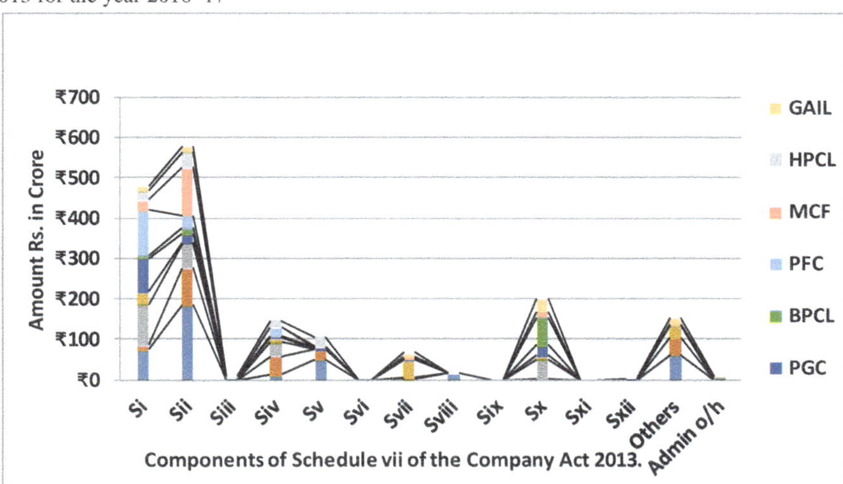

Source Compilation done by the researcher

The average number of **CPSEs spending as per Schedule VII** of the Company Act 2013 was 05. S-i and S-ii were the most preferred item of spending by the CPSEs from the Schedule VII and were preferred by all the 10 CPSEs. S-vi was the least preferred item of spending by the CPSE's, preferred by PFC & PGF. Similarly, the items which were not considered at all by any of the CPSEs for their CSR spending for the year were S-ix and S-xi.

During the year 2017–18, as per Table 5.15 above, the **collective CSR spending** done by the selected 10 CPSEs was found to be ₹1999.05 crore, which was found to be spent on the 13 different items of the Schedule VII of the Company Act 2013. While the average actual CSR spending was found to be ₹153.77 crore, the highest CSR spending ₹606.63 crore done towards S-ii, while the lowest CSR spending of ₹1.00 crore was done towards item S-xi of Schedule VII of the Company Act 2013.

It is observed that out of the total 14 items of the Schedule VII, the collective CSR spending was found to be done on 13 items, namely S-i, S-ii, S-iii, S-iv, S-v, S-vi, S-vii, S-viii, S-x, S-xi, S-xii and Others & Administrative Overheads. Similarly, No CSR spending was found to done on 01 item, namely S-ix.

The average number of CSR activities undertaken for **CSR spending by the CPSEs** from the available fourteen items of Schedule VII of the Company Act 2013 was 07. The highest numbers of CSR activities undertaken was 10, by PGC and the least numbers of CSR activities undertaken was 04, by PFC.

The average number of **CPSEs spending as per Schedule VII** of the Company Act 2013 was 05. S-i and S-ii were the most preferred item of spending by the CPSEs from the Schedule VII and were preferred by all the 10 CPSEs. S-xii was the least preferred item of spending by the CPSEs, PFC. Similarly, the items which were not considered at all by any of the CPSEs for their CSR spending for the year was S-ix.

Table 5.15 CSR spending by the selected 10 CPSEs towards schedule VII of the company act 2013 for the year 2017–18

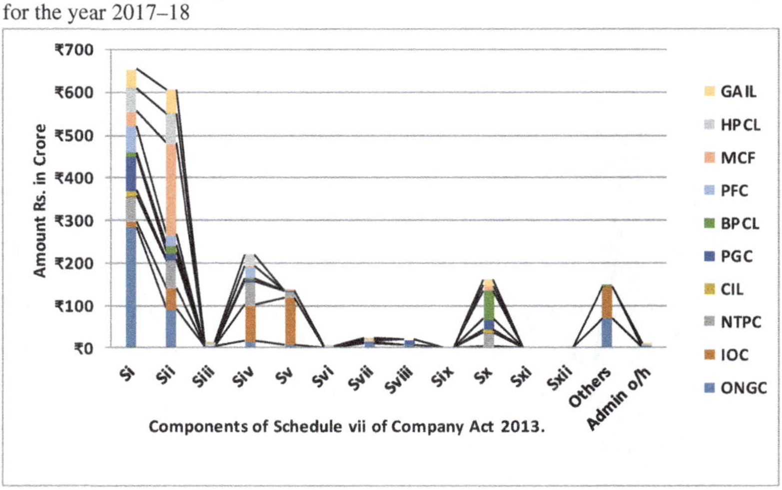

Source Compilation done by the researcher

During the year 2018–19, as per Table 5.16 above, the **collective CSR spending** done by the selected 10 CPSEs was found to be ₹2271.04 crore, which was actually found to be spent on the 12 different items of the Schedule VII of the Company Act 2013. While the average actual CSR spending was found to be ₹174.04 crore, the highest CSR spending ₹611.82 crore done towards "Others", while the lowest CSR spending of ₹0.75 crore was done towards item S-vi of Schedule VII of the Company Act 2013.

It is observed that out of the total 14 items of the Schedule VII, the collective CSR spending was found to be done on 12 items, namely S-i, S-ii, S-iii, S-iv, S-v, S-vi, S-vii, S-x, S-xi, S-xii and Others & Administrative Overheads. Similarly, no CSR spending was found to done on two items, namely S-viii and S-ix.

The average number of CSR activities undertaken for **CSR spending by the CPSEs** from the available fourteen items of Schedule VII of the Company Act 2013 was 07. The highest numbers of CSR activities undertaken was 09, by PGC and the least numbers of CSR activities undertaken was 06, by IOC, BPCL and PFC.

The average number of **CPSEs spending as per Schedule VII** of the Company Act 2013 was 05. S-i was the most preferred item of spending by the CPSEs from the Schedule VII and was preferred by all the 10 CPSEs. S-vi and S-xi were the least preferred item of spending by the CPSEs, namely HPCL and GAIL. Similarly, the items which were not considered at all by any of the CPSEs for their CSR spending for the year were S-viii and S-ix.

Table 5.16 CSR spending by the selected 10 CPSEs towards schedule VII of the company act 2013 for the year 2018–19

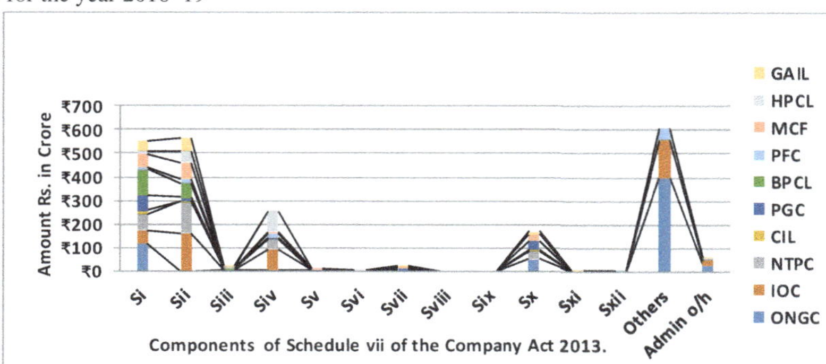

Source Compilation done by the researcher

5.6.8 Inter-CPSEs Analysis of Over/Underspending of CSR Towards Schedule VII of the Company Act 2013

Under this heading, we are analysing the position of CSR spending according to the budgeted CSR allocation to understand the overspendings or underspendings of CSR done by the selected 10 CPSEs towards Schedule VII of the Company Act 2013. Year-wise analysis for has been done collectively and individually, for a duration of five years, starting from the year 2014–15 to 2018–19.

During the Year 2014–15, as per Table 5.17 above, it has been observed that while the net over/under CSR spending done by the selected 10 **CPSEs collectively** was amounting to ₹1055.84 crore underspending, and the net average underspending was amounting to ₹95.99 crore.

The highest underspending amount was found to be ₹566.37 crore, and the lowest underspending amount was found to be ₹0.14 crore. Similarly, the highest overspending amount was found to be ₹106.67 crore and the lowest overspending amount was found to be ₹0.96 crore.

As per the 14 items of the **Schedule VII** of the Company Act 2013, it was found that, while on the three items, CSR spending was overspent. On eight items, CSR spending was underspent, and on the three items, there were no CSR spending at all.

The three items of CSR overspending were S-iv, S-vii and Administrative Overheads. The eight items of CSR underspending were S-i, S-ii, S-iii, S-v, S-viii, S-x, S-xi and "Others". The three items of no CSR spending were S-vi, S-ix and S-xi of the Schedule VII of the Company Act 2013.

Table 5.17 Over/underspending on CSR by the selected 10 CPSEs towards schedule VII of the company act 2013 for the year 2014–15

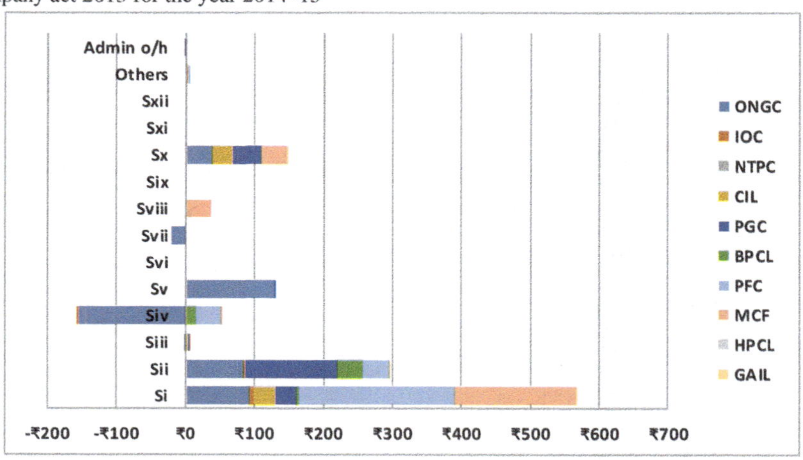

Source Compilation done by the researcher

As per the **selected 10 CPSEs**, PGC was found to have done the highest numbers of CSR underspending on 08 items out of 14 items of the Schedule VII of the company act 2013 and the lowest numbers of CSR underspending on three items was found to be done by GAIL.

Similarly, ONGC was found to have done the highest number of CSR overspending on 03 items out of 14 items of the Schedule VII of the Company Act 2013 and the lowest number CSR Overspending of 01 each was found to be done by IOC and PGC.

HPCL is the only CPSEs who had done their CSR spending strictly according to their CSR budgetary allocations, and therefore, HPCL does not have any overspending, underspending or no spending of CSR for the year.

During the Year 2015–16, as per Table 5.18 above, it has been observed that the net over/under CSR spending done by the selected 10 CPSEs collectively was amounting to ₹641.59 crore underspending and the net average underspending was amounting to ₹58.33 crore. The highest underspending amount was found to be ₹281.04 crore, and the lowest underspending amount was found to be ₹0.17 crore. Similarly, the highest overspending amount was found to be ₹12.57 crore and the lowest overspending amount was found to be ₹0.22 crore.

As per the 14 items of the **Schedule VII** of the Company Act 2013, it was found that, while on the two items, CSR spending was overspent, on nine items CSR spending were underspent, and on three items, there were no CSR spending at all.

The two items of CSR overspending were S-vii and S-xi. The nine items of CSR underspending were S-i, S-ii, S-iii, S-iv, S-v, S-viii, S-x and Others & Administration

Table 5.18 Over/underspending on CSR by the selected 10 CPSEs towards schedule VII of the company act 2013 for the year 2015–16

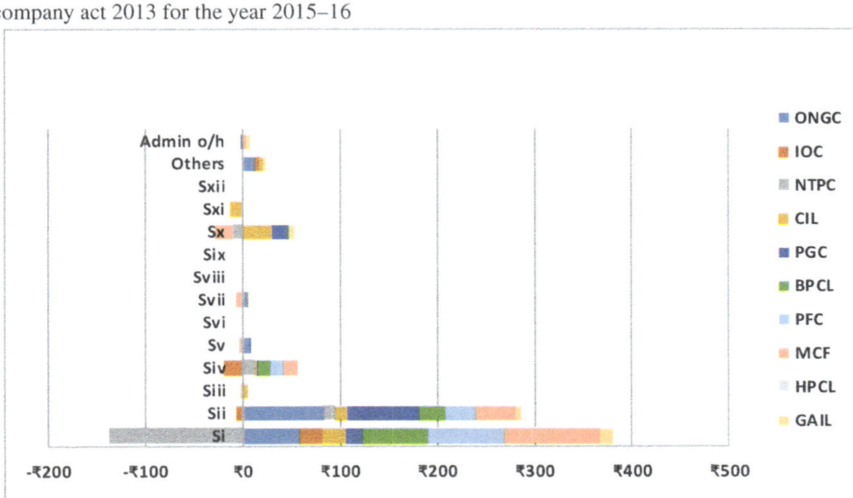

Source Compilation done by the researcher

Overheads. The three items of no CSR spending were S-vi, S-ix and S-xii of the Schedule VII of the Company Act 2013.

Out of the 10 Selected CPSEs, ONGC was found to have done the highest numbers of CSR underspending on 11 items out of 14 items of the Schedule VII of the Company Act 2013 and the lowest numbers of CSR underspending on three items was found to be done by IOC and NTPC.

Similarly, MCF was found to have done the highest number of CSR overspending on 04 items out of 14 items of the Schedule VII of the Company Act 2013 and the lowest number CSR overspending of 01 each was found to be done by PGC.

HPCL is the only CPSEs who had done their CSR spending strictly according to their CSR budgetary allocations, and therefore, HPCL does not have any overspending, underspending or no spending of CSR for the year.

During the Year 2016–17, as per Table 5.19 above, it has been observed that the net over/under CSR spending done by the selected 10 **CPSEs collectively** was amounting to ₹1895.06 crore underspending and the net average underspending was amounting to ₹172.28 crore. The highest underspending amount was found to be ₹965.78 crore, and the lowest underspending amount was found to be ₹1.61 crore. Similarly, only one overspending amount was found to be ₹4.86 crore.

As per the 14 items of the **Schedule VII** of the Company Act 2013, it was found that, while on the 01 item, CSR spending was overspent, on 10 items CSR spending was underspent, and on three items, there were no CSR spending at all.

The only one item of CSR overspending was found to be administration overhead. The ten items of CSR underspending were S-i, S-ii, S-iii, S-iv, S-v, S-vi, S-vii, S-viii, S-x and Other. The three items of no CSR spending were S-ix, S-xi and S-xii of the Schedule VII of the Company Act 2013.

Out of the **selected 10 CPSEs**, three CPSEs, namely ONGC, CIL and PGC Ltd., were found to have done the highest numbers of CSR underspending on 07 items out of 14 items of the Schedule VII of the Company Act 2013 and the lowest numbers of CSR underspending on two items was found to be done by NTPC.

Similarly, NTPC was found to have done the highest number of CSR overspending on 04 items out of 14 items of the Schedule VII of the Company Act 2013 and the lowest number CSR overspending of 01 each was found to be done by 04 CPSEs, namely ONGC, CIL, PGC and BPCL.

HPCL is the only CPSEs who had done their CSR spending strictly according to their CSR budgetary allocations, and therefore, HPCL does not have any overspending, underspending or no spending of CSR for the year.

During the Year 2017–18, as per Table 5.20 above, it has been observed that the net over/under CSR spending done by the selected 10 CPSEs collectively was amounting to ₹910.95 crore underspending and the net average underspending amounts was ₹138.57 crore. The highest underspending amount was found to be ₹910.95 crore, and the lowest underspending amount was found to be ₹0.42 crore. Similarly, the highest overspending amount was found to be ₹1.00 crore and the lowest overspending amount was found to be ₹0.06 crore.

Table 5.19 Over/underspending on CSR by the selected 10 CPSEs towards schedule VII of the company act 2013 for the year 2016–17

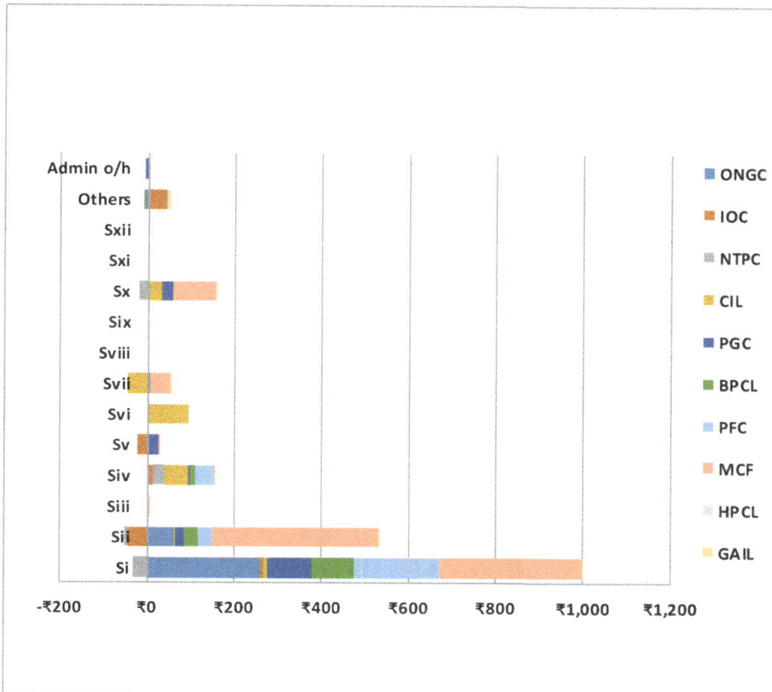

Source Compilation done by the researcher

As per the 14 items of the **Schedule VII** of the Company Act 2013, it was found that while on the two items CSR spending was overspent, on ten items CSR spending was underspent, and on two items, there were no CSR spending at all.

The two items of CSR overspending were S-xi and Administration Overheads. The ten items of CSR underspending were S-i, S-ii, S-iii, S-iv, S-v, S-vi, S-vii, S-viii, S-x and Others The two items of no CSR spending were S-ix and S-xii of the Schedule VII of the company act 2013.

Out of the **selected 10 CPSEs**, ONGC was found to have done the highest numbers of CSR underspending on 09 items out of 14 items of the Schedule VII of the Company Act 2013 and the lowest numbers of CSR underspending on three items was found to be done by PFC.

Similarly, GAIL was found to have done the highest number of CSR overspending on 08 items out of 14 items of the Schedule VII of the Company Act 2013 and the lowest number CSR overspending of 03 each was found to be done by PFC.

HPCL is the only CPSEs who had done their CSR spending strictly according to their CSR budgetary allocations, and therefore, HPCL does not have any overspending, underspending or no spending of CSR for the year.

Table 5.20 Over/underspending on CSR by the selected 10 CPSEs towards schedule VII of the company act 2013 for the year 2017–18

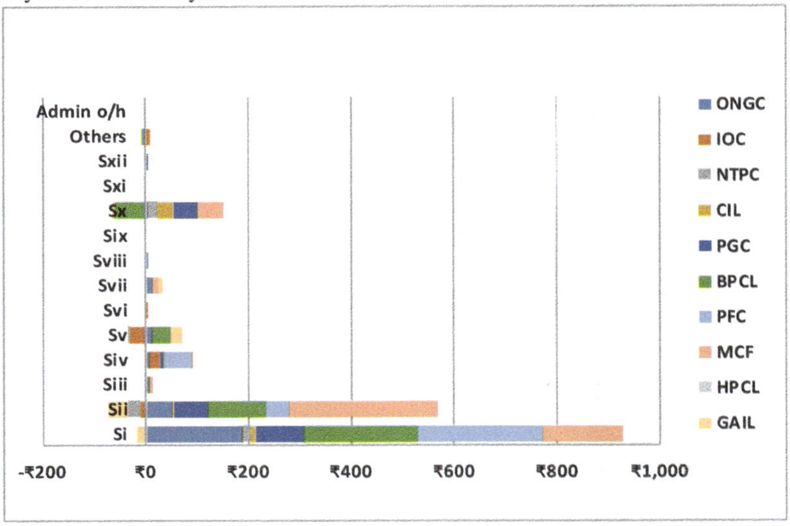

Source Compilation done by the researcher

During the Year 2018–19, as per Table 5.21 above, it has been observed that the net over/under CSR spending done by the selected 10 CPSEs collectively was amounting to ₹2053.25 crore underspending and the net average underspending amounts was ₹205.32 crore. The highest underspending amount was found to be ₹716.45 crore, and the lowest underspending amount was found to be ₹2.89 crore. Similarly, the only overspending found to be done was ₹26.38 crore.

As per the 14 items of the **Schedule VII** of the Company Act 2013, it was found that while on one item CSR spending was overspent, on nine items, CSR spending was underspent, and on four items, there were no CSR spending at all.

While only 01 item of CSR overspending was found to be Administration Overheads. The nine items of CSR underspending were S-i, S-ii, S-iii, S-iv, S-v, S-vii, S-x, S-xii and Others. The four items of no CSR spending were S-vi, S-viii, S-ix and S-xii of the Schedule VII of the Company Act 2013.

Out of the **selected 10 CPSEs**, PGC was found to have done the highest numbers of CSR underspending on 08 items out of 14 items of the Schedule VII of the Company Act 2013 and the lowest numbers of CSR underspending on three items was found to be done by IOC.

Similarly, IOC was found to have done the highest number of CSR overspending on 03 items out of 14 items of the Schedule VII of the Company Act 2013 and the lowest number CSR Overspending of 01 each was found to be done by NTPC & GAIL.

Table 5.21 Over/underspending on CSR by the selected 10 CPSEs towards schedule VII of the company act 2013 for the year 2018–19

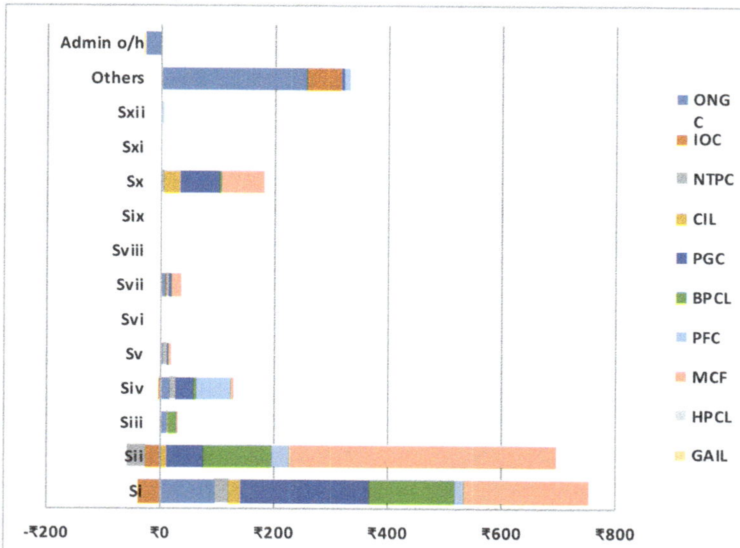

Source Compilation done by the researcher

HPCL is the only CPSEs who had done their CSR spending strictly according to their CSR budgetary allocations and therefore, HPCL does not have any overspending, underspending or no spending of CSR for the year.

5.7 Findings and Conclusion

Based on the interpretation and analysis of the data, the following conclusions have been drawn.

i. It is concluded that the selected 10 CPSEs could utilise only 55% of their available consolidated CSR budget towards their consolidated CSR initiatives. Forty-five percentage of the consolidated CSR budget remained unutilised during the five-year period of study starting from 2014–15 to 2018–19.
ii. CSR spending towards S-i item of Schedule VII-(eradicating hunger, poverty and malnutrition, promoting health care including preventive health care and sanitation [including contribution to the Swachh Bharat Kosh set up by the central government for the promotion of sanitation] and making available safe drinking water), has been the most common and most preferred CSR initiative undertaken consistently by all the selected 10 CPSEs throughout the five-year period of the study.

iii. S-ix item of Schedule VII—(contributions or funds provided to technology incubators located within academic institutions which are approved by the central govt.), has not been undertaken by any one of the selected 10 CPSEs towards their CSR initiative throughout the five-year period of the study.
iv. All the selected 10 CPSEs under the study have done CSR spending beyond the specified 12 items of Schedule VII of the Company Act 2013. One such item has been mentioned as "Others" while another as "Administrative Overheads". Therefore, it is concluded that there is urgent need to expand Schedule VII of the Company Act 2013, so that items mentioned in "Others" can be appropriately identified, accounted and reported accordingly.
v. According to rule 4(6) of CSR policy, rules 2014,[15] salaries paid to regular CSR staff as well as employees, who render their services towards CSR operations, are permitted as Administrative Overheads subject to the condition that such administrative overheads must not exceed 5% of the total CSR spending. Most of the CSPEs have either not done any spending under this head or have done so within the prescribed limit of 5% of the total CSR spending, except for ONGC, which had spent ₹27.79 crore, resulting 7.56% of the total CSR spending, during the year 2018–19.
vi. The selected top 10 CPSEs under the study have not been able to utilise their allocation of budgeted CSR funds totally towards their respective CSR initiatives. The reasons for the same need to be extensively investigated to understand and analyse the prominent causes for not being able to utilise the vast pool of valuable resources made available due to Section 135 of Company Act 2013 for fulfilling their corporate social responsibilities.
vii. The selected 10 CPSEs under the study have shown a consistent underspending on CSR activities has been found to be done by all the 10 selected CPSEs. The total budgeted CSR allocation done by all the selected 10 CPSEs was ₹16,019.88 crore, and the actual CSR spending done by all the selected 10 CPSE's was ₹8799.85 crore (55%). There was underspending on CSR initiatives amounting to ₹7220.30 crore (45%) during the five-year period of study starting from 1014–15 to 2018–19.
viii. Among the selected 10 CPSEs under the study, the highest underspending on CSR amounting to ₹2,053.25 crore (28.44%) was found in the year 2018–19 and the lowest underspending on CSR amounting to ₹630.21 crore (8.73%) was found in the year 2015–16.
ix. HPCL is the only CPSEs among the selected 10 CSPEs, who has done its CSR Spending strictly in accordance with their CSR budgetary provisions throughout the five-year period of the study. Thus, there are no cases of over or underspending on CSR by it. While the remaining nine CPSEs under the study have either underspent or overspent on their different CSR initiatives during the five-year period from 2014–15 to 2018–19.
x. The highest amount of CSR spending done towards an item of Schedule VII of the Company Act 2013 found to be done towards "Others" amounting to

[15] http://www.mca.gov.in/.

₹400.70 crore, which was done by ONGC during the year 2018–19. Similarly, the lowest amount" of CSR spending done towards an item of Schedule VII of the Company Act 2013 amounting to ₹0.02 crore each was found to be done towards S-v (protection of national heritage, art and culture including restoration of buildings and sites of historical importance and works of art; setting up public libraries; promotion and development of traditional art and handicrafts) by PGC during 2014–15 and "Others" done by PFC during the 2015–16.

xi. The highest amount of overspending of CSR done towards an item of Schedule VII of the Company Act 2013 amounting to ₹156.52 crore was found to be done towards S-iv (ensuring environmental sustainability, ecological balance, protection of flora and fauna, animal welfare, agroforestry, conservation of natural resources and maintaining quality of soil, air and water [including contribution to the Clean Ganga Fund set up by the central government for rejuvenation of river Ganga]) by ONGC during the year 2014–15. Similarly, the lowest amount of overspending of CSR done towards an item of Schedule VII of the Company Act 2013 amounting to ₹0.04 crore was found to be done towards "Administrative Overheads" by CIL during the year 2014–15.

xii. The highest amount of underspending of CSR done towards an item of Schedule VII of the Company Act 2013 amounting to ₹469.08 crore was found to be done towards S-ii (promoting education, including special education and employment enhancing vocation skills especially among children, women, elderly and the differently abled and livelihood enhancement projects.) by MCF during the year 2018–19. Similarly, the lowest amount of underspending of CSR done towards an item of Schedule VII of the Company Act 2013 amounting to ₹0.02 crore was found to be done towards "Others" by PFC during the year 2015–16.

References

Kadyan, J. S. (2014). Growth and development of corporate social responsibility in India. *Journal of Applied Management and Investment*. *3*(2), 83–87. http://www.jami.org.ua/abstracts3-2.htm

Kadyan, J. S. (2016). Impact of companies act 2013 on the functioning of corporate social responsibility in India. *Primax International Journal of Commerce and Management Research*, *4*(1), 19–23. https://www.primaxijcmr.com/images/PRIMAX%20IJCMR%20VOLUME%20No.4,%20ISSUE%20No.%201%20APRIL%20-JUNE%20-2016.pdf

Kadyan, J. S. (2017). Analysing and understanding the different theories and policies of CSR-corporate social responsibility. *Global Journal of Multidisciplinary Studies*, *3*(5), 259–264. https://doi.org/10.2139/ssrn.3375981

Kadyan, J. S. (2020). Review of CSR-corporate social responsibility initiatives of the top three CPSE's-central public sector enterprises of India. In D. Crowther & S. Seifi (Eds.), *Governance and sustainability: Approaches to global sustainability, markets, and governance* (pp. 257–280). Springer Singapore. https://doi.org/10.1007/978-981-15-6370-6_13

Kadyan, J. S., (2021). Corporate social responsibility in India. In S. Seifi (Eds.), *The Palgrave handbook of corporate social responsibility* (pp. 1–31). Palgrave Macmillan. https://doi.org/10.1007/978-3-030-22438-7_86-1

Ministry of Corporate Affairs. (2020). Indian companies (CSR policy) rules (2014). Retrieved from https://www.mca.gov.in/Ministry/pdf/CompaniesActNotification2_2014.pdf

Mohan, A. (2001). Corporate citizenship: Perspectives from India. *The Journal of Corporate Citizenship, 2*, 107–117. http://www.jstor.org/stable/jcorpciti.2.107

Srivastava, A. K., Negi, G., Mishra, V., & Pandey, S. (2012). Corporate social responsibility: A case study of TATA group. *Journal of Business and Management, 3*, 17–27. https://doi.org/10.9790/487X-0351727

The Companies (Corporate Social Responsibility Policy) Amendment Rules. (2020). Retrieved from https://www.mca.gov.in/Ministry/pdf/csr_26082020.pdf

The Institute of Company Secretaries of India. (2019). The Companies (Amendment) act (2019). A perspective. Retrieved from https://www.icsi.edu/media/filer_public/9e/cf/9ecf2ae7-9df1-45f8-9653-b1a4975d32ba/recent_changes_in_companies_act.pdf

Part III
Reading Fraud by CSR, AI and Social Entrepreneurship

Chapter 6
Adherence of Fraud Pentagon Dimensions in Cases Reported by Security Exchange Commission in United States Between 2018 and 2019

Ana Lucia Fontes de Souza Vasconcelos, Liliane Cristina Segura, Mariana Alves Serbonchini, Natany Kristine dos Santos Silva, Paulo Andre Matta Chagas, and M. Abu Naser

Abstract The business scenario has presented, over the last few years, several cases of accounting scandals, many of which have been taken forward by organs' inspection agencies of their respective countries. Nevertheless, several companies have seen an increase in their experience with fraud, according to research conducted by PWC (Fraud, who´s committing it? Who knows about it? https://pwc.blogs.com/files/2009-uk-survey-report.pdf, 2018). Many authors discussed the fact that pressure is one of the most recurrent characteristics when analyzing fraud cases. Literature presents, among several other theories, the fraud triangle, the fraud diamond and the pentagon fraud model developed by Crowe. The pentagon of fraud model names five factors that cause fraud: pressure, opportunity, rationalization, capability and arrogance. The overall objective of the study is to verify the adherence of the fraud pentagon theory to the complaint processes carried out by the United States Security Exchange Commission (SEC) between 2018 and 2019. To meet the purposes of the research, this study has as specific objectives the survey of the files of fraud reporting processes from 2018 to 2019, seeking to identify the dimensions of the fraud pentagon theory; analyze the dimensions present in the situations that preceded the case raised by the SEC, according to its judgment reports; and identify which were the most prominent dimensions in the cases reported by the SEC. Thus, the identified cases of fraud and corruption that were disclosed on the SEC website in the period 2018–2019 were identified and analyzed through the fraud theory and the framing of the motivations in the fraud pentagon (on the qualitative nature of the fraud), in order to describe, based on the theoretical assumptions of the fraud pentagon, the dimensions

A. L. F. de Souza Vasconcelos · L. C. Segura (✉) · M. A. Serbonchini · N. K. dos Santos Silva · P. A. M. Chagas
Universidade Presbiteriana Mackenzie, São Paulo, Brazil
e-mail: Liliane.segura@mackenzie.br

A. L. F. de Souza Vasconcelos
e-mail: ana.vasconcelos@mackenzie.br

M. A. Naser
London Metropolitan University, London, UK

M. A. Naser (ed.), *New Approaches to CSR, Sustainability and Accountability, Volume IV*, Accounting, Finance, Sustainability, Governance & Fraud: Theory and Application, https://doi.org/10.1007/978-981-16-9499-8_6

of fraud in the cases raised on the SEC website from 2018 to 2019, in the North American environment, since the information was easier to access. The motivations and purposes of the fraud were analyzed in order to fit them in the dimension or dimensions that best describe the cases analyzed, based on a qualitative descriptive analysis of the information obtained from the documents "SEC Complaint"; it was possible to conclude that the dimensions of the pentagon are still valid and observable in cases reported by the United States tax authorities.

Keywords Fraud · SEC · Fraud · Pentagon · Dimensions · Corruption

6.1 Introduction

In the last decades, there have been several cases of accounting scandals in the business scenario, many of which have been taken forward by inspection bodies of their respective countries (Silva et al., 2012). Nevertheless, several companies have seen an increase in their experiences with fraud, according to research conducted by PWC (2018). Between 2001 and 2002, several cases of accounting fraud were reported by the media, having as main means of action: (i) profit inflation (Toshiba, HealthSouth Corporation, Satyam Computer Services, Tyco International and Dynergy); (ii) the use of accounting gaps (Enron Corporation and Lehman Brothers); and (iii) asset value inflation in WorldCom (Bergamini Junior, 2002; Contábil, 2017).

Thus, it was possible to observe that the fraud in the financial statements of companies, regardless of their size, affects not only the corporation, but the entire chain of relationships with its employees, investors and government entities. The succession of events caused by fraud is one of the main concerns of the North American inspection agencies since accounting manipulations can bring significant losses to the society that actively invests in the stock market (Silva et al., 2012).

In this sense, people with or without prior knowledge of the stock market end up involved in fraud schemes conducted by fake companies, as was the case with *Instaprim Pharmaceuticas* (Appendix). In the case of the United States, where the majority of the population invests a significant part of their savings in the stock market, the losses of fraudulent schemes can lead people to bankruptcy or end up generating a crisis in the business system. In the case of corporations, when fraud schemes are triggered, they suffer from the loss of credibility in the market, both for shares and investments (Bergamini Junior, 2002).

In addition to the concerns of regulatory bodies, public tolerance for inappropriate behavior by corporations or related persons has decreased over time, so that companies and their leaders have been judged for their behavior (PWC, 2018).

According to (Silva et al., 2012), Brazil does not have a structured database containing information about accounting frauds in the country which is easy to access and quickly identified. On the other hand, United States has the Securities and

Exchange Commission (SEC) which provide a database containing the denouncement processes against companies for fraud in the most diverse sectors, such as accounting fraud, use of privileged information, document fraud, among others.

In a study conducted by Condé et al. (2015), which took into account the K-mart fraud case and applied the concept of the fraud triangle and corporate scandals, concluded that all the characteristics of the triangle were present in the case of fraud committed by the company in question, with a main focus on pressure, since they sought to hide the real financial condition of the company.

The theory of the fraud triangle is revisited by several authors, seeking to add greater viability to the characteristics proposed by Cressey in 1950. Schuchter and Levi (2013) proposed that all elements can be changed to fit the difficulties of companies, since a culture of bullying and ambition on the part of stakeholders, can lead to environments in which the propensity to fraud is higher.

Many authors discussed the fact that pressure is one of the most recurrent characteristics when analyzing cases of fraud, Lister (2007) points to three types of pressure: (i) personnel to pay for a certain lifestyle, (ii) employment, due to reward systems according to the operation and (iii) financial return expected by the stakeholders. Albrecht et al. (2008) believe that fraud can occur with or without financial incentives or pressures from behind, valuing the impact behind rationalization, since it depends on the fraudster's perception and the reasons why it comes to justify your actions.

In the initial analyses of Wolfe and Hermanson (2004) raised question about the individual's ability to perform fraud, adding that the functions performed by him can create or exploit the propensity for fraud to occur, pointing out that if the person does not possessed one particular talent, it would not be possible for the fraudulent action to occur. For Vona (2008), on the other hand, the act of practicing fraud is associated with corporate pressures on the individual, so that his position in the entity can lead to the implementation of the action.

Marks (2012), complemented the diamond theory of fraud with a fifth dimension, *arrogance*, or lack of awareness, and it is an attitude of those who believe they are exempt from internal controls, that is, those to whom company policy it would not apply, since he is in a position superior to all others and because he is the person who must take risks. So that this individual becomes more predisposed to commit fraud.

Regarding the application of concepts about fraud, Silva et al. (2012) carried out a study taking into account manipulations in order to deceive external users, with a focus on earnings management. In Brazil, we found that most of the fraud crime was intended to divert resources, evade taxes and remunerate executives, while in the United States the focus was on hiding the real situation of companies, in addition to embezzlement and tax evasion.

As explained above, the intrinsic dimensions of fraud are a relevant component in the discovery and analysis of fraud cases that have arisen over the years. Thus, the research problem takes into account the application of the dimensions of the fraud pentagon in the analysis of the complaint cases raised by the SEC from 2018 to 2019.

Given the above, this research seeks to answer the following question: what is the adherence of the dimensions of the fraud pentagon (rationalization, pressure, opportunity, capacity and willingness to risk—arrogance and ego) in the complaints made by the SEC from 2018 to 2019?

The general objective of the study is to verify the adherence of the fraud pentagon theory to the complaint processes carried out by the SEC between 2018 and 2019. In order to meet the research purposes, this study has as specific objectives to raise the files of the processes on the SEC website of fraud reporting from 2018 to 2019, seeking to identify the dimensions of the fraud pentagon theory; analyze the dimensions present in the situations that preceded the case raised by the SEC, according to its judgment reports; and identify which dimensions were most prominent in the cases reported by the SEC.

With the increase in the practice of management accounting, managers seek to balance the interests of all parties, in addition to assessing the entity's results. However, what ends up happening is an increase in the result for the owners, by hiding the real situation of the company, if the reality of the company is not what is expected by its stakeholders (Nepomuceno, 2002).

Many of the cases analyzed by the SEC concern companies that, in order to maintain the entity's financial appearances, as a way of not being under pressure from investors, used fraudulent means to achieve such results. As was the case observed at Logitech (Appendix), in which the CFO (Chief Financial Officer) and the controller engaged in the fraudulent action of inflating operating revenues in order to demonstrate good results by the company to external users.

In addition to this, other cases such as the use of the fraudster's ability to deceive external users have also been tried by the SEC, and these vary in purpose and method applied. In the case of 4D Circle (Appendix), the company sold investment securities on the pretext of offering more profitable interest rates, in addition to holdings in other companies in the group, also earning as there were sales made by the company. Behind this scheme, the company used improper accounting techniques and misappropriated the amounts invested by investors.

According to research results by Silva et al. (2012), reflections and discussions on what is systematic in fraud cases are necessary, in addition to studying their characteristics, in order to formulate new analysis models.

Authors such as Schuchter and Levi (2013) reinforce the idea that the elements of the fraud triangle can be modified, in order to suit the scenario of the company and the fraudster. Since the financial incentive is not always the main factor that leads to the occurrence of fraud, it can result from internal or external pressure factors.

According to Bayou and Reinstein (2001), it is important to differentiate the fraud object from the fraudulent action, since the action is the intentional activity of seeking its own benefits, while the object is the focus of the action. From this differentiation, it is possible to detect or deter the act of occurring, since they occur by omission or delegation of the activity.

Therefore, since the heterogeneity of the environment increases the uncertainty created, the possibility of fraud cases also increases, contributing to the pressure on the organization, since there is the search for the maintenance of results or the improvement of it (Baucus, 1994).

Therefore, analyzing cases of fraud is important to determine and seek ways to combat fraud, since when you understand the incentives and the means to carry out fraud, it becomes easier to combat it within companies.

6.2 Theoretical Framework

6.2.1 Fraud Theory

For Murcia and Carvalho (2009), it can be understood that fraud is not an "error" or an unintentional action, but an act with the intention of obtaining its own benefit to the detriment of the environment in which it finds itself. Thus, the intentionality of the action is fundamental in determining the situation of fraud.

For Wells (2017), fraud can be presented in 3 different ways: (i) corruption: which is the use of influence or power to obtain one's own advantage and also includes cases of bribery, conflict of interest, economic extortion and discounts inappropriate; (ii) misappropriation: the diversion and misuse of the organization's assets and resources; and (iii) fraudulent statements: it concerns information disclosed by the company that does not represent the reality of the company and its purpose is to induce the user to error.

As shown in the below, the occurrences can be observed according to their origins and type of action involved, highlighting the issue of intentionality of the process (Table 6.1).

In this way, determining what is fraud goes beyond concepts and aspects, but the intentionality of the action developed by the individual in conjunction with the other characteristics is maintained.

Table 6.1 Classification of occurrences

Type of occurrences	Action designation	Source of occurrence
Error	Not intention	Technical fail
Fraud	Intentional, illegitimate and willful action	Obtaining illegal or illicit benefits
Irregularity	Intentional act	Technical ignorance (subject to sanctions)

Fonte Adapted from Lourenço and Sarmento (2008)

6.2.2 Dimensions Used in Fraud Analysis

In 1950, Donald Cressey, an American criminologist, was the one who first studied what motivated people to commit acts of fraud and thus violate the trust placed in them (Machado, 2015). Continuing his research, Cressey mentions that the opportunity to commit fraud arises when the fraudster perceives a way to use his position as a way to solve his problems, since he perceives himself as "untouchable" (Condé et al., 2015; Kassem & Higson, 2012).

Consequently, for Cressey, fraud is not just a behavior that caused some subjective crime, but a situation in which the individual accepted a certain position in good faith and violated the trust of his employer by committing a certain crime (Machado, 2015).

Based on the idea that financial incentives alone would not be enough, Cressey designated the following three elements as necessary, albeit with different impacts, for fraud to occur: (i) pressure, which can be internal or external to the individual, making it the desperate; (ii) opportunity, which may come from the fact that the individual has a position of trust; and (iii) rationalization, which is the way in which the individual seeks to justify the fraudulent action (Condé et al., 2015; Schuchter & Levi, 2013).

The pressure addressed by the author considers the context in which the individual is located. According to Arendt (2008), it is the circumstances of the moment that will influence his decision to perform a certain action, regardless of his morality or character. This action can start in situations of financial difficulties or for social and political reasons, in which the individual cannot present his failure, either because of his personal position, reputation or ego understanding (Alves, 2005; Santos, 2016).

As a result of the situational pressure, a study by PWC (2018) pointed out that the main factors that contribute to the pressure are: financial goals that are difficult to achieve, fear of losing jobs and remuneration dependent on financial performance.

In the case of rationalization, the ethical dilemmas and the individual's discernment must be taken into account, since their moral perception is tested when the fraud is operationalized.

Finally, the opportunity is the perception of how fragile the object of interest is and / or the means by which the fraud will be carried out. In this way, environmental factors are decisive for the creation of opportunities, in the same way that regulated systems, with internal controls, governance and compliance help to minimize the occurrence of the opportunity to commit fraud (Baucus, 1994).

Today, the theory of the fraud triangle is used as a fundamental basis for the development and categorization of risk factors, even though it has a greater focus on people and not on organizations (Schuchter & Levi, 2013).

Continuing the research, according to Wolfe and Hermanson (2004), fraud does not occur without the right person, with the right capacities being involved. This is because only an individual with the capacity to identify a particular gap as an opportunity could take advantage of it. In this way, the diamond theory of fraud goes beyond the opportunity to commit a fraudulent act in terms of environmental or situational factors.

The diamond theory of fraud complements the concept of pressure with the incentive derived from the individual's personal ability. Thus, it is no use for the fraudster to have access to the system he intends to defraud, if he does not have the competence to execute the fraud, since the other characteristics would not be the only ones necessary for the fraud to occur Santos (2016).

According to a study by Abdullahi and Mansor (2015), there are four focal points of capacity that must be observed in order to recognize potential fraud environments:

a) Capability: the individual's need to have certain traits or abilities to commit fraud, at which point the fraudster perceives the opportunity to make the fraud real;
b) Position and function: the degree of hierarchy in the organization in which it is located, with this characteristic that can assist in the mastery of fraud and breach of organizational trust;
c) Intelligence, creativity and ego: in the practice of fraudulent actions, individuals are usually people who understand and are able to exploit the weaknesses of the internal control systems, in addition to using their position to not be detected. In addition, such individuals have great confidence that they will not be detected or believe that they would be able to get out of the situation easily if they were caught;
d) Coercion, deceit and stress: when combined, these three characteristics can help the individual to convince others to commit the fraud or to ignore the fact that the fraud is taking place.

According to Galvis-Castañeda (2017), the condition of knowledge and skills would be part of the opportunity raised by Cressey, since there is an interaction between conditions external to the individual and his cognitive processes, in addition to the need to know how to explore the situation to carry out the fraud. According to Christian (2019), the dimensions of the fraud diamond have a significant effect on the occurrence of fraud.

In the case of the dimensions of the pentagon, for Marks (2012), the fifth dimension would be the manager's arrogance, since the organizations developed in large international operations, with the absence of direct owners of the business and an administration made by directors and managers; thus it is not possible to maintain a great separation between business and personal life of individuals who are on the front lines of companies.

In this context, the leader of the company would have the following characteristics: (i) high ego, (ii) idea that internal controls cannot prevent the fraudulent activities performed by them, (iii) tendency to intimidate smaller positions, (iv) autocratic style of management and (v) fear of losing his position. It would also seek out those who have some similarity in order to maintain the scam.

Due to the fact that fraud is a human behavior, it would be important to understand the factors that can influence the conduct of fraudsters. In this line of thought, Marks (2012) adds that arrogance, or lack of awareness, is an attitude of those who believe they are exempt from internal controls, that is, it would be as if the company's policy did not apply to them, a since it is in a position superior to all others. At this point, the rationalization is also intrinsic since the individual believes his conduct to be acceptable, since he will not be "caught" by internal controls or has the understanding of being "invincible".

As per Galvis-Castañeda (2017) research, the fifth dimension would be the motivation behind fraud, which is defined as the harmonic dynamics between perception, emotion and conduct, making the behavior to be direct and consistent, oriented toward a single objective. And for that, a series of mental processes about the greatest benefit of your conduct is taken into account.

Thus, the fifth dimension takes into account the individual's elevated position in the entity or in its own concept—ego—as well as about the risks and motivations for being willing to commit fraud.

6.3 Methodology

The research developed has a qualitative focus, in which the environment where the fraud occurred will have a significant influence on the factors analyzed, so that the focus is on the process of identifying the characteristics that generated the opportunity for the fraud to occur Godoy (1995). The identification process was done based on the theory of the fraud pentagon.

By seeking an understanding of the phenomenon of fraud applied to selected cases, the research has a descriptive character, in order to establish correlations between the dimensions of fraud and the characteristics raised in the cases analyzed (Freitas & Jabbour, 2011). It is expected that for some processes an additional dimension will be identified.

Its characterization as descriptive can also be done, since it allows the description of the phenomenon of fraud within the context of companies assessed by the SEC. In this context, as the environment is a direct source of data and influences, it was analyzed and computed, taking into account the behavior of those involved in the fraudulent action (Dalfovo et al., 2008; Godoy, 1995).

Since the research does not apply only to a single company, but to the time period from 2018 to 2019, multiple cases will be used, in order to give consistency to the analyzed theme. Thus, it is necessary to have availability and credibility of the information used to characterize the fraud and the characteristics expected for it to occur; therefore, SEC information will be used in a complete manner.

The time period was chosen; since in the last six years, 1800 cases were raised and registered on the SEC website, with the years 2018–2019 having the highest number of cases, with 358 and 321 respectively. Since not all cases registered on the site have the necessary information for this analysis, only those containing documentation (SEC Complaint) were used, resulting in four hundred and thirty-four files in all.

In addition to being a descriptive qualitative study, it will have an exploratory focus, seeking to cross and develop the information raised by the SEC on fraud cases that occurred in the United States based on the dimensions of the fraud pentagon (Freitas & Jabbour, 2011; Voss et al., 2002).

This study delimits geographically the companies that make up the SEC's database, by the composition criterion in the section "Enforcement—Litigation Releases", which presents the cases of litigation related to civil complaint processes carried out by the SEC.

The SEC website provides, in addition to the litigation processes, the accounting and audited information of the companies, so that it is possible to observe quantitatively the amounts involved in the cases of fraud analyzed, although such information is used only as reinforcement for the analysis and not your focus.

For the research, the names of the companies themselves will be used, since the information is public (https://www.sec.gov/litigation/litreleases.shtml), in addition to the keywords, such as: "fraud" and / or "Accounting fraud" on other sites, if a greater level of detail is required.

Some of the cases provide other files such as: "Order", "Final Judgment", "Opinion and Order", "Application", "Memo", "Opinion", "Receivership Order", "Memorandum opinion and order", "Contempt Order" and "Temporary Restraining Order". However, during our analyses, it was found that these documents did not have the information necessary for this analysis, so they were disregarded.

For the analysis of the collected data, a specific typology was developed, classifying accounting fraud according to the dimensions of the fraud and identifying: Company name; Litigant's position; Company branch (more specific); Complaint/case number (SEC Complaint); Complaint and process date; Complaint year; Explanation of the fraud (according to the document); Scam dimension (most prominent) and Scale of dimensions.

Thus, the information described above was collected from the SEC website and synthesized in an Excel table to be categorized according to the prevailing characteristic after analyzing the case according to open complaint. Other information such as the name of the company, the year of the dispute, its sector of activity and a brief description of fraud were also added, in order to enable the assembly of the scale of dimensions (Silva et al., 2012).

Based on the data collected and the previous analysis of the reasons observed in the work of Lima (2019), some of these causes were used and others found in the previous analysis of the cases were added, generating the following table of motivations of the fraud dimensions (Table 6.2).

Based on the causes above and on the reading of complaints, they were considered so that they were then classified according to their presence using the scale of 1–5, being that:

1. It was not observed in the case;
2. The cause of the dimension was observed, with little influence;
3. The cause of the dimension was observed, with some influence, but it was not the determining factor for the occurrence of fraud;
4. The cause of the dimension was observed with a higher level of occurrence, but not yet as the determining factor;
5. The cause of the dimension was observed, and it was the determining factor for the occurrence of fraud.

Table 6.2 Causes related to the dimensions of the pentagon

Dimension	Cause
Pressure	Pressure for result
	Market pressure
	Need to reassure investors/Governance
	Need to raise capital to maintain operations
	Keep up appearances
Opportunity	Control failure
	False control
	Absence of control
	Business knowledge
	Access to information or the person who has the information
	Flexibility of rules
Rationalization	Excess loyalty
	Lack of credibility in compliance
	Omission in the face of fraud
	Security by act
	Knowledge of where and how it could act
Capability	Autonomy
	Knowledge about the business
	Customer knowledge
	Fraud by fraud
	High office in the area or access to high office person
Willingness to risk	Return bet
	High ego
	Low probability of identification
	Daring
	Security through impunity

Source Adapted from Lima (2019)

The classification was carried out according to the understanding that the dimensions need certain circumstances in order to be proven. Scale 1 will be applied when there is no mention of (a) situational or personal pressure, (b) opportunities if due to a lack of control and privileged knowledge, (c) omissions due to rationalization that your actions are not wrong, (d) ability of any kind to perform the action and / or (e) willingness to risk that accompanies the execution.

If these characteristics are mentioned, the importance related to each of the causes of the dimension is observed, in order to classify them from 2 to 5, with 5 being relevant to cases that explicitly present the dimensions aspects of the fraud pentagon.

6.4 Data Analysis

Many of the cases analyzed were related to the North American stock market, in which a large number of individuals carried out fraud based on (i) the sale of shares without having previously registered as a broker dealer or investment consultant at the SEC; (ii) the sale of shares not registered with the SEC; (iii) overvaluation of investment funds; (iv) use of privileged information when buying/selling shares; and (v) appreciation and devaluation ("pump and dump") of the shares of small companies.

However, other cases not directly related to shares were also found, such as the case of Nissan, in which the company's CEO sought to manipulate the compensation paid to him during his years in office. Other cases, such as that related to Volkswagen and "clean diesel", although initially unrelated to the value of the shares, the company's achievements in hiding the real problems of the cars, led to an improvement in the company's market and, as consequence, to the good performance in the market, which affects stocks and other capital in the market—sale of securities (Appendix) (Table 6.3).

According to the data analyzed, it was found that more than half of the fraud cases committed in 2018 and 2019 relate to senior management positions—72% in 2019 and 67% in 2018. The positions vary between company owners, executive heads of different sectors, presidents and secretaries, among others. This fact corroborates the dimensions of the fraud described below, since these are individuals who suffer the pressures imposed by the market and business specialists; recognize when there is an opportunity to act; they have the necessary means so that, through rationalization, they feel beyond any guilt; the ability to carry out fraud; and have the confidence that if they take a risk, they will not be caught.

As it will be analyzed in more detail in the next topics, it was observed during the reading that the dimension that had less causes was pressure, while the one with the greatest presence was the willingness to risk. The others presented causes varying between a determining factor and little/no incidence (Table 6.4).

Table 6.3 Relationship between positions and total cases (2018 e 2019)

Cargo/Posição (Job Position)	Casos Totais (Total Cases)	Casos de Ocorrência (Ocurrency)	Porcentagem (%)
High position (CEO, CFO, president, among others)	434	301	69%
Employee		90	21%
Not related to the firm		43	10%

Table 6.4 Repercussion of causes in the analysis of dimensions

Dimension	Mean		Standard deviation	
	2019	2018	2019	2018
Pressure	1.34	1.13	0.97	0.64
Opportunity	3.22	4.14	1.20	0.84
Rationalization	3.38	4.04	1.50	0.83
Capability	3.79	4.02	1.42	0.76
Willingness to risk	4.46	4.38	0.97	0.61

6.5 Pressure

In 2019, cases involving pressure as the main factor causing fraud were in only 8 of the 208 cases analyzed. In all, the search for good results in companies put pressure on their senior employees to opt for fraud, as a way to reconcile the interests of the market and thus raise more funds.

The number of lawsuits involving pressure as the main cause of fraud remained low in 2018 as well, only 3 cases. However, the characteristics presented remained similar, having to do with the omission of impacts on the results of companies, which could lead to a change in the perception of investors.

As described by Arendt (2008), the circumstances of the moment that influenced the individual's action, regardless of his character or morals. Or according to Santos (2016), for reasons in which the person cannot present his failure. Although in the analyzed years these characteristics have been little presented, they can be confirmed by the analyzed cases.

Some of the cases had their main motivation in situational pressure, as shown by the PWC (2018). This is because the pressure to present good results, achieve an intended remuneration or even not lose a job, were key points in the actions that led to fraud. How was the situation of SeaWorld, which omitted from its presentations the impact of the documentary "Blackfish", which had a negative impact on the company's operations (Appendix).

The case of the College of New Rochelle was a case that presented great pressure points, since the institution started to present a low enrollment rate and, as a consequence, it had little money to maintain its functioning. Thus, one of the controllers observed an opportunity to guarantee this amount through the investment fund that was maintained by the College and began to consume its assets, reconciling this information so that its superiors would not be aware of the situation (Appendix).

Another way of proving the adherence of these cases to the pressure dimension comes from the study by Baucus (1994), who assumes that the need and the pressure, operating together or separately, can lead to an intentional action or not by the individual in order to meet an urgent demand or solve a problem of lack of resources.

6.6 Opportunity

As stated by Santos (2016), the opportunity has to do with the fraudster's understanding of how vulnerable the object is, as well as the means to achieve it. Something that corroborates this understanding is weak internal controls, as well as a lack of monitoring—or understanding that this is minimal—, thus increasing the fascination of fraud Schuchter and Levi (2013).

The opportunity, which was characterized by the knowledge of environmental factors that may influence the occurrence or not of fraud, was observed in 2019 as the main cause of 50 cases, while in 2018 there were 131 occurrences. Unlike the previous dimension, it appeared in different scenarios, related to control (i.e. lack of control when invoicing is collected) or to the knowledge of its sales public (i.e. people of age or without prior knowledge of the capital market).

Most of the frauds are related to the stock market, incorporating in addition to the characteristics mentioned above, also the cases of inside information when trading shares. This is because, in some of the cases, the information was obtained in a casual way, through conversations with high-ranking people within the companies.

Conditions for the opportunity to arise may come from different scenarios, as stated by Baucus (1994). The idea that the SEC would not be a body that would care about investment advisers or unregistered shares is one of the most recurring. Or even with the ambiguity of the laws, which can be proven with the case of Kik Interactive Inc., which believed that its public offering of tokens did not fit into a securities offering (appendix).

In one of the cases analyzed, the owner and managing partner of Spark Trading Group, LLC, a company specialized in trading, took advantage of the contact base he had when managing another company's hedge fund to finance his activities, also taking advantage the lack of knowledge of some of its investors (appendix).

To finalize the confirmation that this dimension can be observed in the cases raised, it is the fact that the individuals believed in the possibility that there was no risk of being held responsible for the fraud, since they had control of the processes, as put by Galvis-Castañeda (2017), as one of the characteristics of the opportunity.

6.7 Rationalization

Complementing the capacity dimension, rationalization comes in line with the individual's discernment of moral values when operationalizing fraud. Based on this dimension, the cases increase to 71 in 2019, in 2018 there was a drop if compared to the opportunity, closing in 112 cases.

As mentioned, the moral perception of when faced with an opportunity will guide the fraudster's action. In other words, it is necessary that he justify his actions to himself, softening his ethical standards and making fraud something "ethical" (Santos, 2016).

It may be noted in some cases in 2018 and 2019 that the litigants acted in bad faith, even though the information received was confidential. As in the case of Slobodan Dragojlovic, an independent software consultant, who made the purchase of shares in the company that his brother worked by discovering inside information about the merger between two companies and making significant gains from this information (appendix).

In other cases, the individual seeks to maintain the image of not being responsible for the fraudulent action, maintaining a self-justification discourse that helps him to overcome the fear of the consequences (Galvis-Castañeda, 2017). Such characteristics are proven in the actions that led to several Ponzi schemes, in which the individual promises high returns on his clients' investments, but makes the payment of these "returns" based on the new investments of other clients, at the same time that it diverts part of the values for its own use.

The adherence of this dimension in the cases raised is corroborated by the studies by Schuchter and Levi (2013) that raised different rationalization techniques used to reduce the moral risks of actions, among them phrases such as "I'm just borrowing and I will return it", "Nobody will get hurt" and "I deserve more", among others. Being the same disseminated by the culture of beautification of status, observed in several Ponzi schemes.

6.8 Capacity

As described by Santos (2016), the ability refers to the individual's ability to operate the system in order to achieve its goals, explaining the fact that it is not only useful to have access to the system, but it is necessary to know how to use it.

Capacity was a key factor for 106 cases registered in 2019 and 106 in 2018. This is because many of the cases depended on the individual's skills, whether it was time to use the lack of control in the company or to command schemes for illicit sale of shares.

To prove this point, the case of Skyworks Solutions, Inc., in which an engineer, who should not have access to financial information of any kind, took advantage of the hours when there were no administrative employees to use the system and gather information in order to that it could trade company shares based on this data (appendix).

The skills of the individual in relation to their knowledge of the business and their customers were a key point in some cases. According to Galvis-Castañeda (2017), these experiences generate greater skills and competences to cause fraud. In some, the fraudster used the knowledge he had about certain communities (Catholic, Jewish or from certain countries) to apply scams with the sale of shares. Or even from the business, when he knew that certain clients—the wealthiest—would not notice the high brokerage fees being paid.

In other cases, individuals used their privileged access to certain information to carry out transactions in the market based on privileged information, as was the case of the former vice president of Ziegenfuse, who, knowing about the merger of the company with another, bought real estate securities and obtained $ 64,000 profit (appendix).

In summary, in the case of capacity, the dimension is proven based on the cases analyzed based on the fact that if the individual did not have the knowledge and, consequently, the ability to understand and explore the deficiencies of the system or the absence of specific rules and/or unknown to certain individuals, the fraudster would not be able to carry out the fraud (Galvis-Castañeda, 2017).

6.9 Risk Provision

For Marks, arrogance, or lack of conscience, is an attitude of those who believe they are exempt from internal controls, that is, it would be as if the company's policy did not apply to them, since they are in a position superior to all others, placing the fifth dimension as something inherent in the actions of individuals.

The cases involving risk disposition were the highest in 2019, with 153 of the 211 analyzed. In 2018, that number was 166 out of 223 cases. This was because many of those investigated were presumed to be out of the risk of being identified or their scheme falling apart. Not only that, but also the boldness to take risks in order to achieve greater returns.

The case of Volkswagen (Appendix), in which the orientation toward a better position in the American market led the company to commit a series of fraudulent arrangements, in order to hide a systemic problem that would cause the loss of the market and a low performance.

This case and many others fit the definition of Marks, as well as that of Galvis-Castañeda (2017), which describes the fifth dimension as the motivation behind fraud, being defined as the harmonic dynamics between perception, emotion and conduct, making the behavior direct and consistent, oriented toward a single objective.

The unique objectives envisaged by the cases ranged from self-enrichment, covering up the company's financial problems, obtaining better results than the real ones, among others.

The disposition to risk was the most recurrent of the five dimensions in all years, because, as stated, the individual who committed the fraud thought he was above any risk or control, making the fraud something possible to be executed and reconciled, if confronted by the authorities or superiors.

6.10 Final Considerations

This study aimed to analyze the characteristics of fraud cases reported by the SEC between 2018 and 2019, to verify the adherence of the dimensions characterized by the Pentagon Theory of Fraud. As pointed out in the analysis, the characteristics described in the SEC's working papers proved to be compatible with those of the theory, making it possible to analyze the causes of the fraud dimensions.

From the qualitative and descriptive analysis of the causes, the results indicated a greater manifestation of cases involving the disposition to risk as the main determinant of the occurrence of fraud, and, in the parallel analysis of positions that most performed fraudulent acts, the top executives were those that prevailed as propitiators.

The case study confirmed that even today the dimensions of the fraud pentagon are present in fraudulent actions. And, by getting to know and clarifying the reasons behind the fraudulent acts, it also becomes easier to prevent and combat them in Brazil and other countries.

As shown by other studies, such as that of Bergamini Junior (2002), despite attempts to reduce the advance of fraud with the adoption of laws such as SOX, the corporate environment still presents structural flaws, such as analysis of employees' behavioral characteristics, the exercise of auditors' technical independence and ethical standards. Sometimes, profit or results lead to decision making that might not be the right one, while different individuals seek to maximize their profits at the expense of the well-being of the company or others.

Since it is still necessary to deepen the studies on the subject, his contribution is linked to the characteristics and cases analyzed, which can assist in the creation of models that explain the causes of fraud in different realities and/or countries. Another reason is to influence the discussion and reflection on possible systematic factors, such as not so strict internal controls and little knowledge about investments, and their impacts on the motivation of fraud.

It is worth mentioning that, when analyzing all cases, questions not proposed for research in this article were observed, such as: (a) the reasons for some companies to cease to exist and others to remain active; and (b) if there were penalties for the companies and people involved.

Finally, through the theory of the fraud pentagon it is possible to conclude that this is still a current topic that requires further research, in order to determine new possible factors that may influence the occurrence of fraud. In addition, the greater the dissemination of information about it, the more individuals will be able to seek to identify and prevent it.

Appendix: List of Cases Available at Security and Exchange Commission (SEC)

Case 1:18-cv-08480. 2018
Seaworld Entertainment, Inc.
https://www.sec.gov/litigation/complaints/2018/comp24272-seaworld.pdf
Access: June/14/2020.

Case 1: 18-cv-1498. 2018
Spark Trading Group, LLC
https://www.sec.gov/litigation/complaints/2018/comp24080.pdf
Access: June/14/2020.

Case 1: 18-cv-4192. 2018
Bryan R. Ziegenfuse
https://www.sec.gov/litigation/complaints/2018/comp24298.pdf
Access: June/14/2020.

Case 1:19-cv-02787. 2019
Keith Borge
https://www.sec.gov/litigation/complaints/2019/comp24436.pdf
Access: June/14/2020.

Case 1:19-cv-05244. 2019
Kik Interactive Inc.
https://www.sec.gov/litigation/complaints/2019/comp-pr2019-87.pdf
Access: June/14/2020.

Case 1:19-cv-08798. 2019
Carlos Ghosn and Gregory L. Kelly
https://www.sec.gov/litigation/complaints/2019/comp24606.pdf
Access: June/14/2020.

Case 2:18-cv-09456. 2018
Slobodan Dragojlovic
https://www.sec.gov/litigation/complaints/2018/comp24338.pdf
Access: June/14/2020.

Case 2:19-cv-02734. 2019
Yuh-yue Chen
https://www.sec.gov/litigation/complaints/2019/comp24454.pdf
Access: June/14/2020.

Case 2:19-cv-13024. 2019
Donald A. Milne III and Instaprin Pharmaceuticals, Inc.
https://www.sec.gov/litigation/complaints/2019/comp24484.pdf
Access: Oct/17/2019.

Case 3:19-cv-01391. 2019
Volkswagen Artiengesellschaft, Martin Winterkorn, Volkswagen Group of America Finance, LLC, and VW Credit Inc.
https://www.sec.gov/litigation/complaints/2019/comp24422.pdf
Access: Oct/17/2019.

Case 3:16-cv-02023. 2016
Erik K Bardman and Jennifer F. Wolf
https://www.sec.gov/litigation/complaints/2016/comp-pr2016-74.pdf
Access: Nov/17/2019.

Case 3:19-cv-01391. 2019
Volkswagen Aktiengesellschaft, Martin Winterkorn, Volkswagen Group of America Finance, LLC, and VW Credit, Inc.
https://www.sec.gov/litigation/complaints/2019/comp24422.pdf
Access: April/06/2020.

Case 4:17-cv-00321-Y. 2018
Richard J. Greenlaw, NECS LLC (F/K/A New England Cannabis Solutions LLC), NYCS LLC, MAINECS LLC, VTCS LLC, MASSCS LLC, NHCS LLC, RICS LLC, CTCS LLC, FLCS LLC, ILCS LLC, IACS LLC, LOUCS LLC, MICS LLC, MNCS LLC, NDCS LLC, NJCS LLC, OHCS LLC, PENNCS LLC, UPCS LLC, and WICS LLC
https://www.sec.gov/litigation/complaints/2018/comp24246.pdf
Access: Nov/17/2019.

References

Abdullahi, R., & Mansor, N. (2015). Fraud triangle theory and fraud diamond theory. Understanding the convergent and divergent for future research. *International Journal of Academic Research in Accounting, Finance and Management Sciences, 5*(4). https://doi.org/10.6007/IJARAFMS/v5-i4/1823

Albrecht, W. S., Albrecht, C., & Albrecht, C. C. (2008). Current trends in fraud and its detection. *Information Security Journal: A Global Perspective, 17*(1), 2–12. https://doi.org/10.1080/19393550801934331

Alves, F. J. dos S. (2005). *Adesão do contabilista ao código de ética da sua profissão: estudo empírico sobre percepções*. Universidade de São Paulo.

Arendt, H. (2008). *Eichman em Jerusalém: Um relato sobre a banalidade do mal*. Companhia das Letras.

Baucus, M. (1994). Pressure, opportunity and predisposition: A multivariate model of corporate illegality. *Journal of Management, 20*(4), 699–721. https://doi.org/10.1016/0149-2063(94)90026-4

Bayou, M. E., & Reinstein, A. (2001). A systemic view of fraud explaining its strategies, anatomy and process. *Critical Prespectives on Accounting, 12*(4), 383–403.

Bergamini Junior, S. (2002). A crise de credibilidade corporativa. *Revista Do BNDES, 9*(8), 33–84. https://web.bndes.gov.br/bib/jspui/bitstream/1408/11576/3/RB%2018%20A%20Crise%20de%20Credibilidade%20Corporativa_P_BD.pdf

Christian, N. (2019). Analysis of fraud triangle, fraud diamond and fraud pentagon theory to detecting corporate fraud in Indonesia. *The International Journal of Business Management and Technology, 3*(4).

Condé, R. A. D., Almeida, C. O. F. de, & Quintal, R. S. (2015). Fraude contábil: ANÁLISE empírica à luz dos pressupostos teóricos do triângulo da fraude e dos escândalos corporativos. *Gestão and Regionalidade, 31*(93), 94–108. https://doi.org/10.13037/gr.vol31n93.2929

Contábil, J. (2017). *Os 9 maiores escândalos contábeis do mundo.* FENACON Sistema Sescap/Sescon. http://fenacon.org.br/noticias/os-9-maiores-escandalos-contabeis-do-mundo-2609/

Dalfovo, M. S., Lana, R. A., & Silveira, A. (2008). Métodos quantitativos e qualitativos: Um resgate teórico. *Revista Interdisciplinar Cientifica Aplicada, 2*(4), 01–13.

Freitas, W. R. S., & Jabbour, C. J. C. (2011). Utilizando estudo de caso(s) como estratégia de pesquisa qualitativa: boas práticas e sugestões. *Estudo and Debate, 18*(2), 7–22.

Galvis-Castañeda, I. E. (2017). Geometria del Fraude. *Cuadernos de Contabilidad, 18*(45), 74–85.

Godoy, A. S. (1995). Introdução à pesquisa qualitativa e suas possibilidades: Uma revisão histórica dos principais autores e obras que refletem esta metodologia de pesquisa em Ciências Sociais. *Revista de Administração de Empresas*, 57–63. https://doi.org/10.1590/S0034-759019950002 00008

Kassem, R., & Higson, A. (2012). The new fraud triangle model. *Journal of Emerging Trends in Economics and Management Sciences (JETEMS), 3*(2), 57–63.

Lima, D. S. (2019). *O uso do pentágono da fraude no processo de identificação de riscos de corrupção em empresas privadas.* Universidade Presbiteriana Mackenzie.

Lister, L. M. (2007). A practical approach to fraud risk. *Internal Auditor*, 61–65. https://search.proquest.com/docview/202735734?accountid=12217

Lourenço, M., & Sarmento, M. (2008). A fraude contabilística e o ambiente empresarial. *TOC*, 34–37. https://www.occ.pt/downloads/files/1224596074_34a37_contabilidade.pdf

Machado, M. R. R. (2015). *Investigação da ocorrência de fraudes corporativas em instituições bancárias brasileiras à luz do triângulo de fraude de Cressey.* Universidade de Brasilia (Dinter UnB/UFG.

Marks, J. (2012). *The mind of fraudsters crime: Key behavioral and environmental elements.* Fraud Conference. https://www.fraudconference.com/uploadedFiles/Fraud_Conference/Content/Course-Materials/presentations/23rd/ppt/10C-Jonathan-Marks.pdf

Murcia, F. D.-R., & Carvalho, L. N. (2009). Conjecturas Acerca do Gerenciamento de Lucros, Republicação das Demonstrações Contábeis e Fraude Contábil. *Contabilidade Vista & Revista, 18*(4), 61–82. https://revistas.face.ufmg.br/index.php/contabilidadevistaerevista/article/view/340

Nepomuceno, V. (2002). *A queda da contabilidade gerencial e a ascensão da fraude contábil no Estados Unidos.* https://www.researchgate.net/publication/273866700_A_queda_da_contabilidade_gerencial_e_a_ascensao_da_fraude_contabil_nos_Estados_Unidos

PWC. (2018). *Fraud, who´s committing it? Who knows about it?* https://pwc.blogs.com/files/2009-uk-survey-report.pdf

Santos, R. A. (2016). *Modelo preditivo de fraude ocupacional nas organizações privadas.* Pontificia Universidade Católica de São Paulo (PUC/SP).

Schuchter, A., & Levi, M. (2013). The fraud triangle revisited. *Security Journal, 29*(2), 107–121.

Silva, A. H. C. e, Sancovschi, M., Cardozo, J. S. de S., & Condé, R. A. D. (2012). Teoria dos escândalos corporativos: Uma análise comparativa de casos brasileiros e norte-americanos. *Revista de Contabilidade Do Mestrado Em Ciências Contábeis Da UERJ (Online), 2*(4), 01–13. https://doi.org/10.12979/rcmccuerj.v17i1.5406

Vona, L. W. (2008). *Fraud risk assessment: Building a fraud audit program* (1st ed.). John Wiley and Sons. https://www.wiley.com/en-br/Fraud+Risk+Assessment:+Building+a+Fraud+Audit+Program-p-9780470129456

Voss, C., Tsikriktsis, N., & Frohlich, M. (2002). Case research in operations management. *International Journal of Operations and Production Management, 22*(2), 195–219. https://doi.org/10.1108/01443570210414329

Wells, J. T. (2017). *Corporate fraud handbook: Prevention and detection* (5th ed.). Wiley.

Wolfe, D. T., & Hermanson, D. R. (2004). The fraud diamond: Considering the four elements of fraud. *CPA Journal*, 38–42. https://digitalcommons.kennesaw.edu/cgi/viewcontent.cgi?article=2546&context=facpubs

Chapter 7
Artificial Intelligence (AI) and Big Data: Impediments to Fraud in Social Entrepreneurship Using Participative Social Governance

M. Abu Naser and Mumtahina Zaman

Abstract Here we present some reflections and a series of observations on the position of AI and Big Data in the management and implementation of social entrepreneurship. As major corporations, social entrepreneurs support the community by providing jobs and revenues. AI has the potential to support significantly, and in diverse ways, the work of people who are engaged in social entrepreneurship. The objective of this study is to bring to light the fraudulence occurring in social entrepreneurship and the necessary steps to eliminate it. The finding of this study indicates three examples of fraud occurred in social entrepreneurship. This study argues that the role of AI and Big Data is more effective in fraud prevention rather than manual investigation. These types of fraud could be prevented by participation of groups in decision making and monitoring process known as participative social governance. Therefore, the purpose of this paper is to discuss types of fraud risks in social enterprises in Bangladesh and the effective ways to prevent it. This study used case study in the social enterprises in Bangladesh, to be specific in Dhaka, the capital of Bangladesh. This study argues that the role of AI and Big Data is more effective in fraud detection rather than manual monitoring of personal. The study will be useful for scholars, policy makers and regulators who are interested in social entrepreneurship in Bangladesh. The review provides guidance to researchers and educators and offers directions and suggestions for practical study in this field. Overcoming the obstacles, entrepreneurship researchers rise to the challenges and engage in analyzation of the probable and concrete AI and Big Data applications. It includes both increased partnerships with other sectors (e.g., computer science and information systems) and updated study frameworks including different forms of technology inside entrepreneurship research (e.g., modern testing centers and exposure to the newest and efficient innovations or networking networks and social network like seminars, meetings, data depository).

M. A. Naser (✉)
London Metropolitan University, Holloway Road, London, UK
e-mail: m.naser1@londonmet.ac.uk

M. Zaman
BRAC University, Mohakhali, Dhaka, Bangladesh
e-mail: mumtahina.zaman@g.bracu.ac.bd

M. A. Naser (ed.), *New Approaches to CSR, Sustainability and Accountability, Volume IV*, Accounting, Finance, Sustainability, Governance & Fraud: Theory and Application, https://doi.org/10.1007/978-981-16-9499-8_7

7.1 Introduction

There is no denying the significance of AI and Big Data in the field of social entrepreneurship when identifying fraud. The influence of artificial intelligence (AI) and Big Data has gained growing attention and concern in a multitude of science and technology fields over the last few years, however, there has been minimal evaluation in contemporary entrepreneurship research (Obschonka & Audretsch, 2019). There are some risks of fraud occurrences in social entrepreneurship. In 2018, the author in his article stated the potential for AI and Big Data is an effective tool for entrepreneurs with no background in technology. Human agents may be provided with extreme type of help while having a critical context that allocates entrepreneur data knowledge. He supported as the advice that can be effectively and easily delivered to the particular clients. Safety of the public includes relatively similar reasoning for applying AI and Big Data. It can be produced due to the potential that withholds along with the emergency of staff inside its dispatchers. The life-saving data can support quick decision taking with productivity according to its ability (Dineshkumar Soni, 2018).

In 2019, Obschonka and Audretsch mentioned that the integration of Big Data processing as part of a wider approach for fraud detection by artificial intelligence (AI) allows modern technology to identify sophisticated and increasingly complicated fraud incidents efficiently and reliably. Moreover, at the same time it is not surprising that we have not yet seen unprecedented and unparalleled developments in the field. In other terms, AI and Big Data growing have an effect not just on methods, but also on the goal that is being researched using such strategies in entrepreneurship study. As such, the science environment and the reality may co-evolve, or even innovate, into an AI-infused research and practice conglomerate. This means that AI will progressively minimize the gaps between entrepreneurship study and research, with experience moving from analysis to implementation and vice versa (and even blurring barriers between science and practice) as they rely on entrepreneurship scholars, researchers, students and professionals to train proactively for potential scenarios (Obschonka & Audretsch, 2019).

7.2 Literature Review

7.2.1 AI

Artificial intelligence (AI) is widely recognized as an algorithm method to learn and analyze. Learning is characterized as the skill to update an algorithm's coefficients and parameters so that the relationship within input and output data can be recognized. AI algorithm is represented in a number of statistical equations, including big data, artificial neural network, image analysis and so on (Jarrahi, 2018). Data analysis by AI is a technique of machine learning that moves input information across several layers (typically three or more) to examine output. One among AI's most compelling features that could have disrupted job opportunities is the capacity to interpret content, identify trends and improve decision making than humans while being motivated throughout many situations to conduct tasks. Throughout banking, AI was widely used to investigate trends and anticipate major developments (Davenport, 2016).

For example, financial distress estimation has typically been performed using a standard equation that uses financial ratios to mathematically determine the probability of this occurrence. These simplified calculations are associated with a high degree of errors, given market volatility. To measure distress more precisely, Bae suggested the use of function vector supporting radial base function machines (RSVM) and correlated its observations with other data analysis tests. The report concluded that a more reliable prediction of financial hardship could help with improved CFO and boardroom decision taking. It concluded that this latest insight could assist with improved business assessment, which would help investors (Bae, 2012).

The nonlinearity and sophistication of financial details to use AI to forecast stock prices are valued and are efficient was suggested. This method can be highly accurate and enable corporate governance to gain financially, counteract crises, increase productivity and generate profits. An analysis of more than 60 papers written between 1990 and 1996 found that the artificial neural networks (ANNs) were applied primarily to forecasts of recession and stock prices. This is widely addressed that ANN includes data access, and finance is one of the most important fields for successful use of ANN. A more recent research reviewed papers written between 1994 and 2015 to classify the use in industry of neural networks. They found that most neural network implementations were only in forecasts of recession and market price but also in decision making and mission classification (Wong & Selvi, 1998).

7.2.2 Big Data

Big Data is a function that represents the massive amount of data that saturates an enterprise on a regular basis (both structured and unstructured) ("Big Data—Definition, Importance, Examples and Tools", 2020). But what's relevant isn't the volume of information. It is what organizations are dealing with the related data. Big Data can be mined for observations leading to stronger business policy and strategic steps. The corporations' use of Big Data is increasingly popular these days to overperform their peers. Both existing players and new entrants in most sectors will use the approaches arising from the evaluated data to perform, evolve and gain interest. Big Data allows corporations develop potential opportunities and brand separate product areas that can integrate and interpret data from the market. Such businesses have adequate product and service knowledge, customers and providers, customer expectations that can be collected and evaluated (Obschonka & Audretsch, 2019). While the term "Big Data" is relatively new, the act of gathering and storing large amounts of information for eventual analysis is ages old. The concept gained momentum in the early 2000s when industry analyst Doug Laney articulated the now-mainstream definition of Big Data as the four Vs: Volume, Velocity, Variety and Veracity. Big Data is neither just how much data an organization has nor how an organization uses the data it gathers. Every organization uses data in its own way; the more successfully a business uses its data, the greater the ability to expand. The company can collect and analyze data from any source to find answers ("Big Data—Definition, Importance, Examples and Tools", 2020).

The latest results of Big Data in both the government and industry can be considered. Including tailored advertisement, schooling and vast sectors (healthcare, insurance, manufacturing or banking) that have already been listed, through real-life situations, in guest services or entertainment. By 2020, 1.7 megabytes of data will be created every second for every human on the planet, and there is tremendous opportunity for data-driven organizational development in the hospitality field ("Big Data—Definition, Importance, Examples & Tools", 2020).

Big Data in governance among many other fields will have a great impact—local, regional and global. For so many complicated problems on the agenda now, politicians are throwing through their job trying to make sense of all the information they are collecting and taking important choices that concern millions. Governments, whether it be any country, emerge face-to -face on an almost continuous basis with a huge amount of data. Reasonably, they have to keep a record of the numerous public records and repositories. The comprehensive analysis and evaluation of this data assists in endless ways for the governments ("Big Data—Definition, Importance, Examples and Tools", 2020).

7.2.3 Social Entrepreneurship

Social entrepreneurship incorporates a social mission's intensity with a picture of business-like professionalism, creativity and commitment. Most social entrepreneurs are introducing entirely new projects incorporating creative and sometimes risk-taking strategies to build sustainable solutions, and concocting new goods and services. Others are entering global mutual businesses associated with their values and passions. Entrepreneurs generate new organizations that can be established as charitable or conventional enterprises, with the overall objective of achieving large-scale, sustainable and comprehensive change through entrepreneurial companies. Social entrepreneurship produces globally to deliver innovative solutions with specific social and environmental impacts (Obschonka & Audretsch, 2019).

Ever since it emerged in the 1990s the concept of social entrepreneurship has been in continuous evolution. However, for today, it is seen as conducting corporation for a social cause by placing over economic performance the general interest of the society. In this way, income is no longer the primary aim of industry in social enterprise but a means of pursuing larger aims related to better society and the environment (Jarrahi, 2018). All such legal economic sectors are the entitled performers of the social and solidarity economy, often established as social purpose corporate entities, low profit limited liability companies, cooperatives or benefit corporations. Social enterprises are naturally managed according to the core values of corporate social responsibility (CSR) and sustainable development. The people who control those social enterprises are called social entrepreneurs. And as we have talked about, their measure of success is not about profit. Instead when their company has somehow changed the planet, they feel effective. Some of the things that could have hindered social entrepreneurship longevity is fraud. Any fraud that occurred in a social project would bring double break-down. Furthermore, there could be severe financial damage, as a result the social enterprise lacks public trust. When social enterprise loses public trust the venture is over, as social entrepreneurship is a trust sector. In fact, businessmen, like sociopreneurs, are rule breakers, i.e., the rule breakers in the good ways would drive creativity in society, but without solid ethics they could break the rules. Social entrepreneurs can bring about significant societal changes.

7.3 Methodology

This study is designed to involve the general population of Dhaka in Bangladesh. The respondents for this survey are chosen randomly from elementary to official levels in the in the city who have some general and basic knowledge regarding the fraudulence in social entrepreneurship. The reason behind such a random survey is that students and job holders each level is a prime component of participating social entrepreneurship. In addition, the respondents in this research are randomly selected because of their particular characteristics or their backgrounds, beliefs or opinions and are in the right decision to provide the necessary knowledge. The research involves the use of statistical data to assist the selected respondents in collecting data.

Senior executives from multiple social organizations were asked to review the questionnaire for appropriate coverage of specific study target measurements. Based on their careers and expertise in the area of fraud detection, they were considered to be professional judges to assess the questions as valid or otherwise. The questionnaire was decided based on their concepts and opinions, and any invalid things were discarded. The survey form includes two parts, including the description of the respondent, and the efficacy of the fraud prevention and identification strategies (6 questions) and the efficacy of the fraud avoidance and identification machine application / software (8 questions).

The respondents were asked to indicate their opinion on the degree of effectiveness of each technique or strategy based on seven-point Likert-type scale, which range from "strongly agree" (scored as 1) to "strongly disagree" (score as 7). The self-administered survey was distributed via social media (i.e., e-mail, Facebook, Messenger, WhatsApp along with a number of social media links). The actual survey was administrated from mid-April until the end of June 2020. A total of 100 responses came out after the period of two and a half month's survey.

The study has extensive information on risks of fraud and strategy implemented to prevent fraud in social enterprises. The data was tabulated using EMES framework of Social Enterprise. Efforts on organizational level of designing and implementing participative governance for fraud prevention program were focused.

"The detailed stages of the study done in four stages are as follows:

1. Preliminary investigation:
   (a) identify the history and background of establishment of Enterprises,
   (b) study the organization structures and governance
2. Observation, interview and focus groups discussion to identify risk of frauds
3. Investigate practice of participatory governance as a major strategy to prevent fraud
4. Triangulation and write-up result. Using multi-source of data to formulate model and key factors of successful fraud prevention program using participatory governance." (Suryanto et al., 2019)

Risks of fraud using risk matrix of consequences and likelihood were measured. The risk using self-understanding of magnitude and likelihood of such risk occurred which were identified and scored. The score is one until three of low, moderate and high. The score of consequences was then multiplied with the score of likelihood as risk score. Based on the result, a confirmation and discussion act were taken places with the enterprises.

7.4 Findings: (Results and Discussions)

Based on the literature review and data collection, the prevalence of fraud in social entrepreneurship into three groups, i.e.,

1. the misappropriation of asset,
2. fraudulent in financial reporting,
3. mission camouflage.

These are the most likely in social entrepreneurship fraudulence occurrences. Risk (in social enterprise management) does exist. For instance, the possibility of misappropriation of properties or misuse of power (in social enterprise) is significant. The vague boundaries between social mission and business mission allowed management the ability to behave not in compliance with the basic principle of internal management. As part of democracy, management may make justification for any wastefulness or misappropriation of money. For instance, several costs are overbudget due to the desire to increase the reach of the recipient or certain anomalies are approved because of the social cohesiveness. Such motives are goods but provide management with loopholes to cover up their unethical practices (Suryanto et al., 2019).

The second category possibility is financial statement fraudulent. Fraudulent financial reporting occurs when management of financial records deliberately undervalued or overstated estimates. Social entrepreneurs must accomplish both financial and social objectives. Revenue results are typically calculated by sales and benefit and social priorities measured by market availability. The management of social enterprise could alter the demonstration of financial and non-financial monitoring to have an appropriate valuation report in order to organize the quality assessment for each panel (Suryanto et al., 2019).

The last set of threats is camouflage project. Mission camouflage intentionally misleads investor or stakeholder that the practices are for social purpose, but in reality, the exercises are for personal or small groups only. A significant danger is the act of deceiving the public into getting source of funds. Several entrepreneurs rush into this approach as social entrepreneurship gains popularity and coverage. For certain if the purpose is genuine, that would be effective, however if it is for "marketing gimmick," it would basically affect the image of social entrepreneurship (Suryanto et al., 2019).

The possibility of cash loss has also been reported, in addition to the three fraudulent events above. Social enterprises tend to be in a restricted window in doing business. When the market fluctuates, their cashflow control may be seriously impacted. Social enterprise management team is also very selective in investing the budget. They try to seek to optimize the use of cash flow for short-term operations, and focus on funding from headquarters for long-term investment. It is a significant analysis because the majority of management in social corporations fail to recognize budgetary management.

The social enterprises in Dhaka implement some internal controls to prevent and minimize such risk to happen:

7.4.1 Internal Controls to Prevent and Minimize Risk to Happen

1. Building public trust through transparency on financial reporting (Most enterprises use third-party web-based accounting software. Public can see the report and detail of transaction of each Unit)
2. Inviting public participation to monitor the performance and compliance of management, since the Board of Management alone would not sufficient to oversee all Unit
3. Preparing business plan to minimize risk of business failure. Before started each business unit, management previously prepares business plan to minimize the risk of bankruptcy
4. Always conducting sound and healthy business practice
5. Preparing monthly and weekly plan as guideline for employee to work

(continued)

(continued)

6. Implementing level of authority of cash disbursement. Head of Business unit only allowed to approve the budget. He/she needs to have approval first from director for spending above the fixed budget
7. Preparing financial and performance reporting and discuss with head quarters and board of management each three months

Source Interview and observation of documents in the Social Enterprises in Dhaka (Suryanto et al., 2019)

The management of social enterprises is attempting to conduct their business in order to improve their market operation with the aid of the new technology. While they do have some Headquarters convenience, this is not enough. They retain results and financial responsibility for each three months and annually. Every year the full annual and monthly budget is presented at the Entrepreneurs' Ceremony. If there are blurred occurrences, each enterprise will be held responsible by the community. Members of the board of directors will have the maximum ability to control if they should endorse the accountability report, approve the strategic initiative and take on the management of social enterprises. This system makes verifying and monitoring method very successful. Community companies have also formally introduced certain new corporate guiding principles, such as remuneration, performance assessment, information systems and participative budgeting and analysis. When interviewed if such methods can be learnt, they responded that they recruited experts and advisors who are recently trained or who have quite a variety of professional expertise to assist them prepare for the documents (Suryanto et al., 2019).

7.5 Conclusion

In addition to enriching and shaping future social entrepreneurship, AI and Big Data may also turn certain facets of the current real-world trends into entrepreneurship. In the recent past, similar historical developments have been noted, such as the emergence of e-commerce or emerging "players" in entrepreneurial financing such as crowdfunding, which have affected the social entrepreneurship analysis agenda. Or, in other words, AI and Big Data could impact the procedures as well as the target being practiced in entrepreneurship using such techniques. As such, the realm of science and the real-world phenomena could co-evolve, or even co-transform, into a research and practice conglomerate infused with AI. This could indicate in further narrowing the distance between study and practice on entrepreneurship through AI, with outflows of expertise from study to practice, and vice versa (or also obscuring the boundary between research and practice) (Obschonka & Audretsch, 2019).

In the past few years, we have experienced a huge rise in the volume of data (text, pictures, audio, video, etc.), both at global and economic levels. Data arrives from all over the world, from various and varied outlets such as contracts, client contacts, contact centers, social media, computers, emails, faxes, etc. The tendency is to use such data for the entity's benefit (designing tactics, finding incentives, building reputation, preventing and detecting fraud, etc.). Using Big Data processing systems and the tools devoted to these activities allows comprehensive and in-depth study of the trends and structures of the informal economy, crime and corruption, as information and communication technology is a sine qua non tool of recorded (formal) economics. While there is a broad range of advanced instruments available on the analytical market capable of helping and improving the anti-fraud operation, the survey findings sadly show that managers are not taking advantage of them. Although these are rather idealistic projections, since only the future can predict whether there will be further transitions or continuity. In fact, it would be important to consider whether AI and Big Data can contribute to a transformation in social entrepreneurship and its study and practice in terms of more or less gradual or revolutionary developments, or if it will potentially cause a full paradigm shift and a "explosion of information" that transforms the principle of the system (Obschonka & Audretsch, 2019).

References

Ashkan Eliasy a, *, Justyna Przychodzen b, The role of AI in capital structure to enhance corporate funding strategies

Bae, J. K. (2012). Predicting financial distress of the South Korean manufacturing industries. *Expert Systems with Applications, 39*(10), 9159–9165.

Big Data—Definition, Importance, Examples and Tools. RDA. (2020). Retrieved from August 27, 2020. https://rd-alliance.org/group/big-data-ig-data-development-ig/wiki/big-data-definition-importance-examples-tools

Dineshkumar Soni, V. (2018). Role of AI in industry in emergency services. Retrieved from 7 May 2020. https://www.researchgate.net/publication/343106692

Davenport, T. H. (2016). Rise of the strategy machines. *MIT Sloan Management Review, 58*(1), 29.

Jarrahi, M. H. (2018). Artificial intelligence and the future of work: Human-AI symbiosis in organizational decision making. *Business Horizons, 61*(4), 577–586.

Liu, L., Wang, Y. (2005). Radial basis function support vector machine based soft-magnetic ring core inspection. In: *International conference on computational and information science*. Springer.

Obschonka, M., & Audretsch, D. (2019). Artificial intelligence and big data in entrepreneurship: A new era has begun. *Small Business Economics, 55*(3), 529–539. https://doi.org/10.1007/s11187-019-00202-4

Omonijo, D., Olowookere, E., & Ogunyemi, S. (2020). Stimulating the role of civil societies for sustainable development through social entrepreneurship. *International Journal of Psychosocial Rehabilitation, 24*,(04), 15. Retrieved from August 22, 2020. https://www.researchgate.net/profile/Dare_Ojo_Omonijo/publication/339826579_Stimulating_The_Role_of_Civil_Societies_for_Sustainable_Development_Through_Social_Entrepreneurship

Suryanto, R., Mohammed, N., & Yaya, R. (2019). Fraud frevention for sustainable social enterprise using participative social governance. In *The first international conference on financial forensics and fraud*, p. 7. Bali; Research gate. Retrieved from January 27, 2020. https://www.researchgate.net/publication/341455293_Fraud_Frevention_for_Sustainable_Social_Enterprise_Using_Participative_Social_Governance.

Wong, B. K., & Selvi, Y. (1998). Neural network applications in finance: A review and analysis of literature (1990–1996). *Inf Manag, 34*(3), 129–139.

Chapter 8
Restorative Justice: Application to Corporate Fraud

Mark Lokanan

Abstract The paper argues for a restorative justice approach to corporate fraud. Drawing from the restorative justice and crime control literature, the paper argues that given the nature of corporate fraud and the stakeholders affected, restorative justice can have an impact in disciplining corporate offenders. Restorative justice is a different framework from criminal justice or regulatory proceedings. A restorative approach to justice sees crime as involving all the stakeholders affected by the offender's action. In the aftermath of a crime, the victim, offender, community, and authorities meet to discuss how the offenses affected the lives of the victims and neutralize its collective impact. The fundamental postulate of restorative justice is to promote healing and rehabilitation for the victim, offender, and all other stakeholders affected by the crime. While restorative justice is widely used in addressing street crimes, little effort has been made to link its applicability to corporate crime.

Keywords Restorative justice · Corporate fraud · Accounting · Crime control · Ethics and laws

8.1 Introduction

In 2005, sixty-three-year-old Canadian businessman and a key player in the sponsorship scandal that brought down the Canadian Liberal party in 2006, Paul Coffin, stood up in a Quebec Superior Court room and apologized "to all Canadians" (Gabbay, 2006, p. 421). In an emotional testimony, Mr. Coffin stated: "I realize what I have done... I know I have tarnished my family name. I know I have disappointed all Canadians" (p. 421). According to the Justice John Gomery's (2005) report, Mr. Coffin was charged with eighteen counts of fraud involving more than $2 million in false invoices billed to Communication Canada. After a guilty plea, to fifteen of the eighteen charges, "Coffin mortgaged his house, cashed out retirement funds, and borrowed $500,000 from friends and family in order to repay the government

M. Lokanan (✉)
Data Scientist, Machine Learning, Royal Roads University, Colwood, Canada
e-mail: Mark.Lokanan@royalroads.ca

M. A. Naser (ed.), *New Approaches to CSR, Sustainability and Accountability, Volume IV*, Accounting, Finance, Sustainability, Governance & Fraud: Theory and Application, https://doi.org/10.1007/978-981-16-9499-8_8

one million dollars" (Gabbay, 2006, p. 421). Mr. Coffin was sentenced to two years of community service and order to take business ethics lessons (pp. 421–422). In a tribunal case, Thomas Clarke ("Clarke") was charged by the Investment Dealers Association of Canada with forgery and unauthorized trading in client's account. As a result of Mr. Clarke's action, his clients' loss was estimated to be around $500,000. In the settlement agreement with the IDA and victims, Mr. Clarke expressed a genuine remorse for his conduct and cooperated with the IDA's Enforcement Staff "to resolve the financial problems arising from his conduct" (IDA *v. Clarke*, 2007: 19). In his settlement agreement, Mr. Clarke admitted to unbecoming conduct and agreed to contribute the sum of $368,000.00 toward restitution for the victims' losses. To do so, Mr. Clarke, "substantially liquidated his financial holdings and mortgaged his interest in his matrimonial residence" (para. 19). Mr. Clarke was asked to pay a fine with terms and conditions attached to his continued employment as an investment advisor. In both these cases, the genuine remorse and reparative actions taken by the respondent worked to mitigate against harsher penalties being imposed by the sentencing judges.

The cases raise many pertinent questions concerning the appropriate response to white-collar crimes (also see Braithwaite, 2013). Here are situations where the offenders confessed to the offenses, showed genuine remorse, apologized to their victims, and most of all, repaired the harm caused by making amends for their misconduct. Considering the monetary impact of the offenses, many have queried whether the punishment meted out in these cases is proportionate to the harm (FAIR Canada, 2011). An argument could be made that while apologies were needed, it could have been use in another forum and not as a justification for the lenient sentences imposed by the judges (Gabbay, 2006, p. 422). A forum that sees the victims, offenders, community, and authority as equal stakeholders in the quest for justice; a forum that first view crime as primarily an offense against human relationships and secondarily a violation of ethics and laws; and a forum that recognizes crime is wrong and should not occur, and when it does occur, there are dangers and opportunities, i.e., to reintegrate the offender back into society (Braithwaite, 2013) (Bazemore, 2005; Bazemore & Mara, 2004; Hayes, 2005; Ivancevich et al., 2008; Lokanan, 2009). One possibility of such a forum is restorative justice. A restorative justice approach to crime involves all the stakeholders affected by the offender's action. In the aftermath of a crime, the victim, offender, community, and authorities meet to discuss how the offenses affected the lives of the victims and neutralize its collective impact (Braithwaite, 2007). The objective is to promote healing and rehabilitation for the victim, offender, and all other stakeholders affected by the crime.

While restorative justice is widely used in addressing street crimes (Braithwaite, 2013; Kuo et al., 2010), little effort has been made to links its applicability to white-collar crimes (Braithwaite, 2013) (Gabbay, 2006). The central research questions are as follows: Is restorative justice approach a reasonable response to corporate accounting fraud? Is restorative justice an appropriate response to corporate accounting fraud? To answer these questions, the paper draws on evidence from prior literature to illustrative that restorative justice is a way forward to address corporate fraud. More specifically, the paper argues that restorative justice intervention is warranted and can be applied to high-profile white-collar crime cases that affect thousands of victims. The mental image I want the readers to project is one of professional accountants, who collude with executives to act fraudulently in order to manipulate their companies' financial statements.

8.2 Focus on Professional Accountants

In this paper, the focus is on corporate accounting fraud, rather than the wider area of corporate malefaction (Braithwaite, 2013; Dellaportas, 2014; McMullan, 1992). Corporate accounting fraud is used as a reference point to include crimes that involves CEOs and their inner circles of accountants, lawyer, and information technology experts. Both the victims and offenders who fall under the category of "corporate fraud" are heterogeneous, ranging from the offender (CEO, accountants, lawyers) to the victims (blue-collar and white-collar worker, poor and elderly, and institutions) all of whom are stakeholders of the fraud (Levi, 2002, pp. 149–151). Analyses of the impact of restorative justice can incorporate the wider reference to which the offenders and victims belong at the time of the offense and at the time sanctions is applied (p. 149).

8.3 "What Works" in the Crime Control Literature

In order to examine "what works" in the crime control literature, it is important to first address what does not work (Braithwaite, 2009; Simpson et al., 2013, p. 234). We know from previous research that "compliance rules imposed and 'policed' by the government with an emphasis on punitive sanctions for violators" does not work because individuals who take part in crimes that involved huge monetary rewards are instrumental actors (Simpson et al., 2013, p. 233). This is because their "decisions and actions flow from a cost–benefit assessment of both the pecuniary and non-pecuniary pros and cons associated with illegal activity" (p. 233). If the benefits of committing the "crime are high and the risk of discovery and punishment is low, then criminal opportunities increase as actors perceive less risk associated with illegal activities" (p. 233). We also know that punitive punishment is not a deterrent to criminality (Braithwaite, 2013; Hartney, 2006). The extant literature on the issue suggests that individuals "who are released from prison have a remarkably high probability of going back to prison" (Braithwaite, 2009, p. 440). So, what are the types of measures that have proven to work in the crime control literature? The identification of "hot spots" has proven to be an effective approach for strategies aimed at crime prevention (Lokanan, 2015). Sherman (1995) stated that the identification of "high-risk spaces, times and groups of people, and targeting smartly attuned preventive and rehabilitative measures at the hot spots" has also proven to be effective crime control measures (as cited in Braithwaite, 2009, p. 440). Early intervention of hot spots and treating identified risks lead to more evidence-based intervention (p.440). We also know from the extant "deterrent literature that improved detection makes more difference than the increased punishment that might or might not be delivered by stronger laws" (Braithwaite, 2009, p. 440; Simpson et al., 2013, pp. 233–235). It might be useful for commentators who are calling for "heavier penalties for white-collar offenders" to pause for a minute and apply "these fairly banal lessons" from the criminology literature to prevent the next major corporate scandal (Braithwaite, 2009: 440; (Gabbay, 2006, p. 436).

8.4 The Limitations of Criminal Prosecution and Punishment

8.4.1 Criminal Justice Administration

Corporate scandals produced a flurry of headlines during 2002, and resulted in notable prosecutions, pleas, and convictions for some of America's largest corporations (Ball, 2009, p. 278; O'Connell, 2004, p. 737; Rezaee, 2005, p. 278).[1] Executives from Enron, Tyco, WorldCom, Quest, Adelphia, and ImClone were all tried and convicted in the criminal justice system.[2] However, while the state has spent millions of dollars on enforcement and has launched an unprecedented effort to prosecute and imprison white-collar offenders guilty of such crimes, substantial evidence suggests that the response remains inadequate (Lehman & Okcabol, 2005; O'Connell, 2004; Sikka, 2009). This inadequacy emanates from a sentence jurisprudence that has historically been driven by the "sensational activities [of]... primarily lower-class crime[s]" (Croall, 1992, p. 4). The result is a trend of leniency for corporate lawbreakers and harsh punitive sanctions for conventional street criminals (Nocera, 2011). As it stands at present, the leniency toward white-collar criminals is mainly because of three factors: enforcement (low visibility), legislation (complexity of cases), and the judiciary (diffusion of victimization).

---

[1] See, e.g., Kathryn Kranhold and Richard B Schmitt, "With Andersen Down, Prosecutors Set Sights on Enron," *Wall Street Journal* C1 (June 17, 2002) (conviction of accounting firm of Arthur Andersen LLP of obstruction charges in connection with Enron matter); Deborah Solomon and Susan Pulliam, "U.S., Pushing WorldCom Case, Indicts Ex-CFO and His Aide," *Wall Street Journal* A1 (Aug. 29, 2002) (detailing charges against WorldCom executives).

[2] Since these crimes undermine the financial system itself, the government is likely to react more vigorously than it would for traditional manufacturing sector crimes.

8.4.2 Enforcement (Low Visibility)

Over the years, the enforcement of corporate crimes has proven to be a difficult task. Law enforcement agencies have to compete with other government departments for scarce resources (Marceau & Mongrain, 2011; Podgor, 2007). Devoting these limited resources to the detection and prosecution of sophisticated corporate crimes is often considered inefficient. This means that law enforcement has to invest its limited resources to fight crimes that are more conspicuous to the public. In the majority of cases, these crimes are conventional street crimes because they are less costly, easier to prosecute and most of all, justify their funding and fulfill political agendas (Croall, 1992, p. 50; Lehman & Okcabol, 2005, pp. 623–627). Local police forces are also too small and ill prepared to tackle the complexities of corporate crime (Coleman, 1987; Lokanan, 2017a). The reason often cited for this phenomenon is that the principal training of police personnel is oriented toward conventional crime; police officers are more likely to be attracted to the more dramatic forms of street crime than to white-collar crime, which typically lacks adrenalin-producing "excitement." ...Because the investigation of [white-collar] crime calls for forms of competence and expertise (e.g., accounting knowledge) that traditional police agencies often lack, the chances of failure, and of being perceived as incompetent, are therefore considerably higher (Friedrichs, 1996, p. 272).

Prosecutors also face insurmountable difficulties to prosecute corporate crime (Hamilton, 2007, p. 2; Strader, 2007, p. 3). Unlike street crimes, where investigation is handled by police officers, in corporate crime cases, investigative duties are handled by the prosecutor's office. Like their law enforcement counterparts, the prosecutor's office is handicapped because they are ill prepared to meet the challenges of corporate crime investigation (Hamilton, 2007; Mitchell et al., 1998). The prosecutor's office is also usually understaffed and lacks the necessary resources to carry out its' mandate (Holtfreter et al., 2008; Rebovich & Kane, 2002). Since prosecutors are evaluated by the number of cases that they prosecute, it is more likely that they will forgo a lengthy corporate crime case for a few street crime cases that they are sure will bring about convictions.[3]

Public perceptions of crime shape the implementation and enforcement of substantive laws in the justice system (Findlay & Hanif, 2012; Holtfreter et al., 2008; Rebovich & Kane, 2002). If there is hue and cry from the public that street crime is causing more harm to society than corporate crime, then laws will be implemented to address such outcries (Costelloe et al., 2002). The perceived risk of victimization increases moral credibility and punitive attitudes toward street crimes (Holtfreter et al., 2008, pp. 51–52). An example of this was the high level of home invasions that occurred in California in the early 1990s. This became a populist issue, which resulted in Republican Governor Peter Wilson introducing the now infamous "Three Strikes

---

[3] It is estimated that on average, a price fixing case takes about twenty-one months to investigate and twenty-three months to litigate. Thus, it is unlikely that a district attorney will spend four years for a white-collar conviction whereas if he/she spends less time and resources on a few street crime cases, his/her output will be more than doubled (See Scott, 1989).

and You are Out Law." As a result, anyone who was convicted for three felonies was sentenced to life. No such fate however is meted out to corporate criminals—even when there is huge media uproar over their devious acts.

8.4.3 Legislation (Complexity of Cases)

There is also the problem of legislative ambiguities. Corporate crime statutes are broad and vague, which makes them difficult to interpret (Lokanan, 2017b). Since corporate crime falls under criminal law in most common law jurisdictions, the government must prove the corporation's intent (*mens rea*) in committing the crime and that the crime did in fact occur (*actus rea*). In addition, the laws dealing with corporate crime have a strong individual bias, which makes it difficult to prosecute corporate criminals. The law not only requires that the executive(s) committed the act, but also that the executive(s) acted with criminal intent. For example, an executive who releases a manipulated financial report and claims that it is accurate has committed a criminal act only if s/he knows that the report is manipulated. This is very difficult to prove since the executive can say that s/he did not know the report was manipulated even though there may be some other executive in the organization's hierarchy who knows about the potential danger of fraudulent financial reporting. Determining both the level of *mens rea* and the physical component of the act are difficult to prove in corporate crime cases (Coleman, 1987, p. 165; Strader, 2007, pp. 4–9).[4]

The role of the judiciary in sentencing corporate criminals is another area of concern. Sentencing goals should be informed by a combination of deterrent, retributive, rehabilitative, or other rationales depending on the severity and type of offense (Grunner, 1997, pp. 630–31). When one looks at the severity of the offense by the level of harm done and the social cost of the crime in determining blameworthiness, it is clear that corporate criminals (in most cases) are given more lenient sentences than street criminals in proportion to the offense (Friedrichs, 1996, pp. 318–321; Holtfreter et al., 2008, p. 50; Gottschalk & Rundmo, 2014, pp. 3–4). A few reasons have been put forth for this outcome. It is argued that because corporate crimes are difficult to detect, an offender who has been caught for the first time is seen as a "first time offender" by the judiciary, even if the individual has been doing the act throughout his or her lifetime (Maddan et al., 2012; Podgor, 2007). This coupled with the fact that corporate criminals are armed with sophisticated defense teams (consisting of top-notch lawyers, expert and character witnesses), works together to predispose the judiciary toward more lenient treatment for the offense (Gottschalk & Rundmo, 2014, p. 3).

[4] This problem is confounded in the United States, since criminal law is primarily adopted and enforced at the state level. As such, each state may have different criteria to gauge these issues. Only a few white-collar crime cases are brought to federal court (See Strader, 2007, pp. 4–9).

Many corporate criminals point to a lifetime of social productivity to minimize their first transgression. In *United States* v. *Paterno*, the judge notes that "the defendants...have lived exemplary lives...They have built stable families and stable businesses. People of distinction and more humble workers in their enterprise write letters or sincere praise, devotion, and appreciation on their behalf. The defendants have not been ungenerous in dealing with friends, employees, family members, and charitable agencies. It may be predicted with reasonable confidence ...that neither defendants will...run afoul with the law again. The fall from untarnished eminence in their communities has been an irreparably grievous blow. The defendants in this frequently heard but still pertinent appeal, 'have been punished enough already.' It is urged earnestly that a sentence of imprisonment would serve ends only of vengeance" (Nagel & Hagen, 1992, p. 1432).[5]

Former Australian Federal Court judge Ray Finkelstein, QC, supported the reasoning offered by the judge in *Paterno*. Judge Finkelstein noted that "most judges believe that the humiliation, loss of job and loss of status experienced by white-collar criminals when they are apprehended, brought to trial and punished is usually sufficient punishment...Judges tended to compromise by imposing short, weekend or suspended sentences...Judges tended to regard imprisonment of white-collar criminals as a 'last resort' because they believed 'imprisonment has a far greater detrimental effect on a white collar criminal'" (Butler, 2012, paras 4–7).

The humiliation of criminal prosecution coupled with the defendants' contribution to civic society serves to mitigate the sentence imposed for their transgressions (Hansen, 2009).

8.4.4 Judiciary (Diffusion of Victimization)

Perhaps the factor that has more of an independent effect in handing out lenient sentences to corporate criminals is a shared value system with the judiciary (Croall, 2001, pp. 131–132; Lynch et al., 2004, pp. 391–392). On some level, "judges are more likely to experience some sense of empathy—apparently more as a function of identification with common values rather than from similar background—with defendants in [corporate] crime cases than in cases involving conventional criminals" (Friedrichs, 1996, 2009, p. 329). Conversely, some judges may give corporate offenders harsh sentences because they are in a privileged position in society and their decision to violate social norms is more reprehensible than the behavior of the impoverished street criminal (Weisburd et al., 2001, p. 237). Judges, who share this view, normally consider corporate criminals more culpable than conventional defendants because they are people with respected and trusted status like themselves (Friedrichs, 2009, p. 329).

---

[5] See the various decisions from the lower courts at *United States v. Paterno*, 375 F. Supp. 647 (S.D.N.Y.), *affd.*, 798 F.2nd 1396 (2d Cir. 1974), *Cert Denied*, 419 U.S. 1106 (1975).

In instances where judges consider corporate offenders more culpable, the defense usually invoke the principle of *stare decisis* and compares prior sentences for similar offense(s). In doing so, the defense either argues for a justification of a sentence other than imprisonment in their own case, or for an appeal and a motion to reduce the sentence initially meted out. For example, in *United States* v. *Browder*, the petitioner challenged the sentence imposed on the grounds that it constitutes "cruel and unusual punishment" and denied the petitioner equal protection under the law:

The basis of the petitioner claim is a study he conducted of 100 cases involving similar white-collar crimes… Of the 100 defendants studied, 20% received fines, probation, or suspended sentences only for acts involving $350,000,000 or more. The other studies received light sentences for a variety of swindles in which the public became victim to members of the Mafia, labor union officials, mayors, attorneys, stock brokers, business executives, bankers, a former state Attorney General, a governor, a federal judge, and others (*United States v. Browder*, 1975).[6]

Similarly, in *United States v. Alton Box Board Co.*, a memorandum was submitted to the Court on behalf of the defendants arguing:

In numerous other price—fixing cases of equal or greater magnitude, some involving more aggravating circumstances than alleged in the indictment, no jail sentences were imposed. …The defendants sought to redraw their pleas of nolo contendere on the grounds that they had relied on their counsel's computations "that in 93% of the antitrust cases in which the defendants played nolo, the resulting fines were minimal and no actual imprisonment resulted" (*United States v. Alton Box Board Co*, 1979).[7]

Thus, even when the judiciary is predisposed to give a sentence that may be proportionate to the actual harm, they are up against strong defense teams that have the resources to argue for a lesser sentence.

Convicted corporate criminals are rarely sent to prison. Once incarcerated, depending on the nature of their crimes, some are sent to minimum security prisons that have often been characterized as "country clubs" or Club Fed" (Friedrichs, 2009, p. 335). Of course, there are those who argue that such a categorization is too general since ImClone's lifestyle doyenne, Marta Stewart and Enron's Jeffrey Skilling were all sent to what was classified as overcrowded prisons, which were not conducive for rehabilitation (Maddan et al., 2012). But these are exception, rather than the rule for corporate criminals. As Friedrichs noted, even though incarceration of corporate criminals is intended to be punitive, it is substantially less punitive than the experience of the typical "street" offenders in maximum-security prisons (Friedrichs, 2009, pp. 329–330). When convicted, corporate criminals are usually sent to half-way houses where they are given banal assignments and are quartered with other inmates in small cubicles or dormitories (p. 335). Depending on the nature

---

[6] For a full discussion see *United States v. Browder*, 398 F. Supp. 1042, 1046 (D. Or. 1975), *affd.*, 544 F.2d 525 (9th Circuit 1976).

[7] See the full details of the memorandum at *United States v. Alton Box Board Co.*, 1979 US Dist. LEXIS 9553 (ND Ill. Sept. 26, 1979).

and severity of their crimes, some are given probation orders in the form of community services (Croall, 2001; Levi, 2002, 2009). Such orders direct the company to publicize its transgressions, or orders forbidding the indemnification of fines.[8]

Two broad generalizations can therefore be made with regards to the CJS and the treatment of corporate criminals. First, while some corporate crime cases end up in court and are given prison sentences, the literature also suggests that they are treated more leniently than street criminals in almost every stage of the CJS. Second, the complexity of corporate crime cases makes it difficult for criminal justice personnel to secure a conviction because they lack the requisite expertise and resources needed to do an effective job. Whether it is this structural failure of the CJS that prevents it from treating corporate criminals the same as street criminals, or whether it is selective bias is not yet known. What is known though is that there needs to be "like treatment for like cases," and in order to do this, there needs to be more research in this contentious area of accounting.

8.5 Effectiveness of Sanctions

8.5.1 Lengthy Prison Terms

Perhaps to make amends for the lenient penalties imposed on corporate offenders, judges in the U.S. have started to impose sentences that are far beyond those received in previous years (Podgor, 2007, p. 731; Strader, 2007, pp. 46–47). For example, Bernard Ebbers ("Ebbers") was sentenced to twenty-five years in prison for his role in the $11 billion accounting fraud at WorldCom's (Crawford, 2005, paras. 1–3); and John Rigas, the founder of Adelphia Communication was sentenced to fifteen years, while his son Timothy Rigas, the Company's CFO, was sentenced to twenty years in prison for their roles in the Adelphia's accounting fraud (Searcey & Yuan, 2005, paras 1–2). Although the courts in handing out these harsh sentences are trying to crack down on corporate fraud, lengthy prison terms may be unnecessary. The crimes committed by corporate criminals clearly injured their victims, but the offenders themselves do not always pose a physical threat to society. So why then is there a need for harsh punitive sanctions for corporate criminals? The argument is that harsh punitive sanction is in order to deter prospective criminal conduct (Podgor, 2007, p. 739). However, despite the harsh sentences that were meted out to Ebbers, Skilling, and the Rigas for example, they did not stop the likes of Bernard Madoff for swindling investors of over $50 billion worth of their pension funds and life savings, nor did harsh sentence serve as a deterrent for Raj Rajaratnam, who was convicted in 2011 of masterminding the biggest hedge-fund insider trading scheme in American history.

---

[8] In a Canadian case, *R v. Bata Industries Ltd.*, the court imposed these orders of probation on Bata for discharging toxic waste into the eco-system. See.*R v. Bata Industries Ltd* (1992), 7 C.E.L.R. (N.S.) 293 (Ont.Prov.Ct.), affirmed (1993), 11 C.E.L.R. (N.S.) 208 (Ont.Gen.Div.).

Corporate criminals are rational actors who objectively weight their options by assessing the risks and rewards associated with their actions (Becker, 1968). Deterrence studies will tell you that to these canny crooks, long prison sentence is counterproductive and only serve to appease public outcry for retribution of their criminal conduct (Levi, 2009; Simpson, 2002; Wright, 2010). Long prison sentence does not send a sufficiently strong and potent deterrence message to the rational would-be corporate offender.

8.5.2 Financial Sanctions

It is difficult to argue that financial sanctions are effective to deter rational and calculating corporate offenders (Ivancevich et al., 2008; Strader, 2007; Weissmann & Block, 2007). The recent GlaxoSmithKline ("GSK") fraud in which the Company misbranded their drugs resulted in a settlement of $3 billion (Thomas & Schmidt, 2012, para. 1), while pharmaceutical giant Pfizer is expected to pay $2.3 billion for deceptive marketing practices and kickbacks schemes (Outterson, 2012, p. 1083). The escalated civil settlement to address the misconduct of GSK and Pfizer begs following question: Are civil penalties which run into billions of dollars enough to deter future corporate misconduct? Former Attorney General of New York, Eliot Spitzer ("Spitzer") does not think so. According to Spitzer, as Attorney General, he sued GSK over similar accusation in 2004 (Thomas & Schmidt, 2012, para. 8). GSK had to pay $87 million to settle allegations that the Company violated the *False Claims Act* by submitting false claim to the U.S. government for underpaying rebates, effectively overcharging a government program (Medicaid), and signed a corporate integrity agreement ("CIA"), lasting until 2008, to "ensure that GSK will accurately reports its 'best price' information to the government in the future" (Wolfe, 2013, p. 7505). In 2005, GSK again paid $150 million to resolve allegations that it violated the *False Claims Act* "through fraudulent drug pricing and marketing of two antiemetic drugs" (p. 7507). Similarly, in 2002, Pfizer paid $49 million to settle allegations that it violated the *False Claims Act*. The U.S. government charged that Pfizer "fraudulently avoided paying fully the rebates owed to the state and federal governments under the national drug Medicaid Rebate program for the cholesterol lowering drug Lipitor [avorvastatin]" (p. 7507). In 2004, a Pfizer subsidiary paid $430 million in fines to settle allegations of civil violations involving illegal off label marketing of gabapentin (Neurontin) and kickbacks (p. 7507). Like GSK, Pfizer signed a CIA which was to last until 2007. A few years later, in 2009 and 2012 respectively, both companies again played guilty to health care frauds in the U.S.

What is evident from these two cases is that large fines or the threat of large fines did not deter corporate malfeasance. As a matter of fact, it seems as if GSK, Pfizer, or for that matter, many other companies are viewing "such fines as merely a cost of doing business—a quite small percentage of their global revenue and often a manageable percentage of the revenue received from the particular product under scrutiny" (Becker, 1968; Outterson, 2012, p. 1083). In 2011, GSK make a profit of

$8.87 billion, while Pfizer made $10 billion, meaning the companies are in a good financial position to take legal risk, despite the fact that fines imposed for corporate fraud has increased manyfold in the last decade (Wolfe, 2013). In imposing hefty fines, the government merely recoups a portion of the financial fruit of firms' past misdeeds (Outterson, 2012, p. 1083). The size of the penalties is a far cry from the benefits associated with of circumventing regulation, and therefore seems ineffective to deter corporate malfeasance.

8.5.3 Recidivism Rates Remains High

Perhaps a better indicator to assess the effectiveness of the penalties imposed on corporate criminals is to look at their recidivism rates. From all accounts, the individuals convicted of corporate crimes are quite often repeat offenders (Perri, 2011; Weisburd et al., 2001; Weissmann & Block, 2007). According to the U.S. Federal Sentencing Commission, even though fraud and larceny offenders have lower recidivism rates [for first time offender]… the recidivism rates for these offenses exceed 50% [for offenders with prior records]…and appear sometimes comparable to the recidivism rates for robbery and firearms (U.S. Sentencing Commission, 2004, p. 13).

These statistics for the most part is conservative and must be interpret with caution. It is worth noting that the recidivism rate for corporate offenders may underrepresent the actual rate at which they offend. As mentioned earlier, corporate fraud prosecutions are notoriously difficult to pursue successfully because they depend on complex financial records and often arcane regulatory schemes, and [corporate] defendants are often represented by skilled and well-financed attorneys. As a result, a "first time" [corporate] offender may have engaged in prior frauds without being detected, charged, and convicted (Weissmann & Block, 2007, p. 286).

A lack of a prior record should not be seen as synonymous with a lack of prior offending (Friedrichs, 2009). A case in point is financier Bernard Madoff ("Madoff"). Madoff was apprehended in 2008 and was sentenced to the maximum 150 years in prison for masterminding a $65 billion Ponzi scheme. Madoff for all his shenanigans, can technically be classified as a first-time offender because this is his first arrest; yet, records have shown that the activities of his fraud extended all the way back to 1998 (Trotman, 2014). Many corporate crimes are a result of months of planning and organization and are not the result of one instance of fraudulent act, which may be more characteristic of a street-level crime (Dorminey et al., 2010, pp. 18–21; Perri, 2011, p. 222).

8.5.4 Financial and Personal Impact

The economic cost to society from corporate crime is staggering (Albanese, 1995; Calavita & Pontell, 1990; Lynch et al., 2004). A study by Kappeler et al., (2000) estimates that in the United States, corporate crimes cost the government anywhere from $174 to $231 billion annually (pp. 121–141). Contrast this with the cost of street crimes, which hovers anywhere between $10 and $13.5 billion each year and the picture is clear—corporate crime inflicts far more preventable violence and economic damage to society than all of the street crimes combined (Lynch et al., 2004). Just to put things into perspective, the annual losses to victims from corporate crime is in excess of $250 billion,[9] compared to estimates of $17.6 billion for losses accrued due to personal and household crimes (Holtfreter et al., 2008, p. 50). These figures indicate that the financial costs of corporate fraud exceeded the cost associated with street crime by thirteen times or more (13.8 times to be exact). There is also another side to the story that often gets excluded from discussions on corporate crime. That side depicts the personal consequences of corporate fraud (Boyd, 2005; Leigh et al., 2000).

Victims must deal with the personal consequences that result from their losses. Boyd's (2005) conducted a study of investment fraud in British Columbia and found that the majority of investors who suffered losses of $50,000 or more, "report extreme or major harm to their emotional well-being, their current financial situation and their retirement security" (Boyd, 2005, p. 28). According to Boyd (2005), "almost 30 per cent of this group report extreme or major harm to physical health, and 20 percent report extreme or major harm to friendships and marital relations" (pp. 28–29). Others highlighted the harmful personal consequences that resulted from fraudulent conduct by corporate offenders (Boyd, 2005; Lokanan, 2014). Yet, while street crimes are being vigorously prosecuted, evidence posits that corporate criminals are allowed to sow their wild oats and continue with what former Chairman of the Federal Reserve of the United States, Alan Greenspan termed their "contagious greed" (Dellaportas, 2013; Podgor, 2007).

8.6 Restorative Justice Theoretical Framework

8.6.1 Definition and Features

Restorative justice programs emerged in the 1980s as an alternative method to deal with crime (Kuo et al., 2010; Lokanan, 2009). According to a restorative justice approach, crime is a violation of people and relationships. It creates violations to make things right. Justice involves the victims, the offender and the community in a search for solution which promotes repair, reconciliation, and reassurance (Zehr, 2015, p. 181).

This definition places emphasis on the harmful effects of the offenders' action and actively involves all the stakeholders in the reparation and rehabilitation process (Braithwaite, 2009, 2012). The emphasis is on crime as a harm that affects people,

[9] It was estimated that adding all other frauds to the $250 billion will raise this total to $400 billion annually (Lynch et al., 2000: 66).

not as law which affects the sovereign state. Crime violates social relationships, both personal and those resulting from being members of communities. Crime is not merely an act of law-breaking; it tears the social or community fabric; it is the violation of one human being by another (Beven et al., 2011, pp. 194–195). Crime is primarily an offense against human relationships and secondarily a violation of the law (Jackson, 2008, p. 217). In this sense, the definition does have a drawback, that is, failure to incorporate crime against the state. As such, I would like to point out that I am not claiming this definition captures the essence of restorative justice; I am claiming that this definition will allow me to highlight and address a common feature of many current restorative paradigms—builds community by providing a constructive framework to guide our responses to crime, conflict, offensive behavior. Restorative Justice is a process to make things as right as possible and includes: attending needs created by the offense such as safety and repair of injuries, relationships, and physical damage resulting from the offense. In developing programs based on restorative justice therefore, these underlying principles should form the foundation for planning, implementation, and evaluation (English, 2013).

8.6.2 *Stigamatic and Reintegrative Shaming*

The principles encompassing the restorative justice approach to wrongdoing have their origins from Braithwaite's (1989) book titled *Crime, Shame and Reintegration*. Its approach is as relevant to professional burglary as to episodic delinquency or corporate crime (Braithwaite, 2009, 2013; Dellaportas, 2014; Ivancevich et al., 2008; Kayes et al., 2007). Braithwaite argues that some societies have higher crime rates than others because of their different processes of shaming wrongdoing (Braithwaite, 2009, 2012). In theory, the effectiveness of restorative justice is based on the principles of reintegrative shaming (Braithwaite, 2012). When shaming is done within a cultural context of respect for the offender, it can be an extraordinarily powerful, efficient, and just form of social control (Dellaportas, 2014). Shaming can be of two types: stigmatic and reintegrative shaming.

8.6.2.1 Stigmatic Shaming

Stigmatic shaming can be counterproductive, making crime problems worse (Dellaportas, 2014). This is referred to as stigmatized shaming (disintegrative shaming), which disintegrates the moral bonds between the offender and the community. The primary importance of stigmatization is that it treats offenders as outcasts and provokes a rebellious and criminal reaction from them: "Shaming that is stigmatizing… makes criminal subcultures more attractive because these are in some sense subcultures which reject the rejecters" (Braithwaite, 1989, p. 102). Disapproval that is stigmatic involves communicating that disapproval with disrespect, where the individuals are labeled with outcast identities (e.g., corporate criminal, tax cheat, fraudster).

8.6.2.2 Reintegrative Shaming

Reintegrative shaming strengthens the moral bonds between the offender and the community. Shaming is reintegrative when it reinforces an offender's membership in the community of law-abiding citizens. This prevents the shamed individual from adopting a deviant master status and is accomplished when shaming maintains bonds of love or respect between the person being shamed and the person doing the shaming. Shaming is directed at the evil of the act rather than the evil of the person, and is delivered in a context of general social approval, which is terminated with gestures or ceremonies of forgiveness (Braithwaite, 1989, pp. 100–1). According to Braithwaite (1989) Reintegrative shaming will work because people are generally deterred from committing crime by two informal forms of social control: fear of social disapproval and conscience (Ashforth & Anand, 2003; Murphy & Dacin, 2011). Punishments or reparation agreements imposed by family members, friends, or other individuals important to an offender are more effective than those imposed by a legal institution (Braithwaite, 1989, p. 34–36). For most people, he argues, fear of being shamed by those they care about is the major deterrent to committing crime because the opinions of family and friends mean more than those of an unknown criminal justice authority (p. 56). Braithwaite (2009) also predicts that restorative may be more effective than traditional courts because they include the direct participation of supporters of both victims and offenders. By including supporters, restorative justice allows wrong doers to be held responsible in the context of a community of care. In such a setting, wrong doers can be held accountable for their acts without being condemned as people. According to reintegrative shaming theorists, this combination of accountability and respect is key to keeping an offender within the community (Braithwaite, 2013).

8.6.3 Evidence on Street Crime

Restorative justice scholars have developed a rich literature on its effectiveness on both street and white-collar crime (Braithwaite, 2012; Goodstein & Butterfield, 2010). Prior evaluative research on restorative justice has mainly focused on violent and property offenders (de Beus & Rodriguez, 2007; Hayes, 2005; Kuo et al., 2010; Rodriguez, 2007). On average, studies on the effectiveness of restorative justice practices have consistently reported a reduction on repeat convictions (de Beus & Rodriguez, 2007; Latimer et al., 2005; Rodrigues, 2007; Sherman et al., 2000). In 1998, the Hudson Institute in the USA carried out the Indianapolis Restorative Justice Experiment to evaluate how effective was group conferencing on first-time juvenile offenders. The study found that offenders in the treatment group had a lower recidivism rate (12.3%) than those offenders in the control group (22.7%).

Latimer et al., (2005) did a meta-analysis of 35 restorative justice program (i.e., 27 victim-offender mediation and 8 group conference programs) from twenty-two studies and found that they were more effective than traditional correction programs in reducing recidivism. In another study, Rodriguez (2007) using official juvenile court data in Maricopa County, Arizona, found that juveniles who participated in a restorative justice program were less likely to recidivate than juveniles in a comparison group. Similarly, de Beus and Rodriguez (2007) found that offenders in the restorative justice program were likely to complete the program and less likely to recidivate than offenders in the comparison group. Upon further examination, de Beus and Rodriguez (2007) found that poverty level had a significant influence on program completion and recidivism. In terms of violent offenses, a randomized Reintegrative Shaming Experiments (RISE) study undertaken in Canberra, Australia the Sherman et al. (2000) found that violent offenders in restorative justice program were less likely to recidivate than violent offenders in the control group. Hayes (2005) reported that that violent young offenders who were part of the Bethlehem, Pennsylvania Restorative Policing Experiment, were less likely to reoffend than offenders who were process through juvenile court.

Restorative justice approaches have been reported to be effective when dealing with serious offenses as well (Kuo et al., 2010). Umbreit and Vos (2000) presented two case studies of capital murder cases involving a victim offender mediation/dialogue session between a surviving family member and the death row inmate facing execution shortly after the mediation session. The five participants (including two offenders and three family members of the victims) felt "a sense of relief" and "peace" and that the intervention had a powerful impact on their lives (Umbreit & Vos, 2000, p. 78). The participants noted that the mediation/dialogue had a powerful impact on their lives and were grateful for the opportunity to meet and engage with the offenders. They recommend that mediation should be done more often.

While restorative justice has had considerable attention in the context of street crimes, a question often raised by scholars is its applicability to white-collar crime (Levi, 2002). This is somewhat ironic given that the theory has its origin on the observations made by Fisse and Braithwaite (1983) in their work on corporate crime (as cited in Murphy & Harris, 2007, p. 902). However, some scholars have suggest that corporate offenders might be insulated to restorative justice approaches, particularly those with pathological or predatory traits who feels no guilt or remorse for his or her actions (Ashford & Anand, 2003; Dorminey et al., 2010; Levi, 2002; Murphy & Dacin, 2011; Murphy, 2012). The debate suggests that understanding the role of restorative justice in curbing corporate fraud is an important policy as well as a theoretical issue (Murphy & Harris, 2007). As a result, a growing number of empirical studies (mostly in Australia) have addressed the usefulness of restorative justice in tax avoidance schemes and compliance with nursing home standards (Ahmed & Braithwaite, 2004, 2005; Grasmick & Bursik, 1990; Makkai & Braithwaite, 1994; Murphy & Harris, 2007). These studies have mostly focused their attention on the role of reintegrative shaming in curbing tax frauds.

Grasmick and Bursik (1990) was the first to examine the potential inhibitory effects of shaming on tax cheats. Using a random sample of 360 participants (respondents), they were asked whether they feel shame to cheat on their taxes and whether they conceive cheating as a problem. The authors found that formal sanction (from the justice system) and the threat of social disapproval from significant others inhibit offenders who for whatever reasons were motivated to cheat on their taxes. Grasmick and Bursik (1990) noted that the anticipated feelings of being shame were found to be a deterrent effect on the decision to offend in the future. Through an analysis of interviews with 30 convicted white-collar offenders, (Benson, 1990) found that offenders were varied in the ways in which they handled shame and anger arising from being stigmatized in court cases. Benson (1990) the perception of being stigmatized by the court process generated anger and rage as well as shame and embarrassment to the participants. Benson (1990) went on to argue that a justice system that is based on disintegrative rather than reintegrative shaming was counterproductive. The findings presented by Benson (1990) are consistent with research that shows shame-related emotions were predicted by perceptions of social disapproval and anger (Harris, 2006) in the context of white-white collar crime.

Indeed, one of the studies that have so far found the strongest support for reintegrative shaming is Makkai and Braithwaite's (1994) work that examines the relationship between the styles of disapproval employed by nursing home inspectors to secure managements' compliance with nursing home standards. Makkai and Braithwaite (1994) did a pre and post-test measures of the compliance record of 410 nursing homes in Australia, for two consecutive years to assess the managers' compliance with nursing home standards. To triangulate the data, they interviewed inspection teams about the regulatory strategies that they used to secure compliance from nursing home managers. The result provides strong support the idea that disapproval that was reintegrative in nature (as opposed to stigmatizing) leads to higher rate of compliance from nursing home managers (Makkai & Braithwaite, 1994, p. 377).

Ahmed and Braithwaite (2004) expanded the research on shame and compliance by investigating the relationship between making additional payments to the state for student loan and child support and compliance with tax law. Drawing on a random sample of 2040 participants, Ahmed and Braithwaite (2004) found that taxpayers who reported that they will displace shame in response for being detected for tax evasion were more likely to report that they have evade their taxes. The authors found that a higher probability of being caught, moral obligation, and possible trustworthiness play significant roles in reducing tax evasion. In another study, consisting of survey data from 477 Australian university graduates Ahmed and Braithwaite (2005) found a significant relationship between the reintegrative measures—shame acknowledgement and shame displacement on tax evasion. Ahmed and Braithwaite (2004) also found that participants who find themselves feeling dissociated from the tax system and have a disregard for the legitimacy of tax law benefited from shaming practices. In a more recent study of 652 tax offenders who were caught and punished by the Australian Tax Office (ATO) for engaging in tax avoidance schemes, Murphy and Harris (2007) found that feelings of reintegration/stigmatization experienced during an enforcement event were related to reoffending behavior. Murphy and Harris (2007) noted that those offenders who had reintegrative enforcement experiences reoffend at lower rates (two years later) than offenders who were not part of the reintegrative enforcement experience. Their findings are consistent with Braithwaite and Braithwaite's (2001) hypothesis that shame-related emotions partially mediate the effect of reintegration on subsequent offending behavior (Murphy & Harris, 2007).

8.7 Responding to the Critics

A number of criticisms have been leveled against restorative justice approaches to crime. The criticisms are mostly philosophical rather than empirical in nature. For ease of presentation, I have grouped what I consider to me the main criticisms of restorative justice under the following headings: soft option and not seen as real punishment; .....I first highlight the criticisms and respond to them (where I can) with rebuttal evidence from empirical sources. I also turn to the fundamental features and values of restorative justice to address the validity of the criticisms. Note that this is not an exhaustive list of the criticisms surrounding restorative justice; rather the criticisms dealt with here were given prominence because of their relevance to the topic (i.e., restorative justice application to accounting fraud) being discussed.

8.7.1 There Are no Victims of White-Collar Crime

In reviewing this critique, I mainly draw upon my own experience as a forensic accountant investigating financial statement and investment frauds. It is commonly claimed that there are no victims of white-collar crime. This claim is based on the argument that there is no direct intent to harm the victims (Croall, 2001; DiMarino & Robertson, 2013). The victims of the Enron or the WorldCom's scandal however would beg to differ. Although producing a fraudulent financial statement is not as violent as being robbed, there are still victims associated with the act. Victimization is corporate accounting fraud is indirect and impersonal and affects stakeholders such as shareholders, creditors, employees, community, and the government. The victims may be difficult to identify because they are usually spreaded out to include a substantial number of people in different geographic locations. The results of a random sample of US citizens with regard to victimization from white-collar crime suggest that victimization is more prevalent than one may be led to believe. Citing the results from the US victimization survey, DiMarino and Robertson (2013) reported that household victimization (45.6%) increased from earlier surveys. The results from the survey also indicated that 36% of respondents had been victimized within the previous year, whereas 62.5% reported being victimized at least once within their lifetime (pp. 11–12). When one does the math, it would appear that 130 million Americans have potentially being victims of white-collar crime once during their lives (p. 12). Fraudulent financial reporting is never a victimless crime and the Enron and WorldCom scandals exemplifies just how many people white-collar crime can directly victimized (Gabbay, 2006, p. 432).

8.8 Applying Restorative Justice to Professional Accountants

This section of the paper takes us back to the work of Dellaportas (2014) and argue that restorative justice is a viable approach to address frauds committed by accountants. By invoking restoratives justice ideals, more particularly reintegrative shaming, it is hoped that it will provide a conduit to secure compliance from corporate actors and steer them away from illegality to legality (Piquero et al., 2008, p. 222). So instead of holding offenders responsible for crimes of the past (passive responsibility), restorative justice seeks to persuade them to take responsibility for putting things right into the future (active responsibility) (Braithwaite, 2013). Victims want to be restored; they want redress and restitution and hopefully an apology from the offender(s). Unlike the criminal justice system which treats the victim(s) as a mere spectator in the process, restorative justice gives all the stakeholders a say in the process.

Figure 8.1 shows a restorative justice process of all the stakeholders involved in the alleged crime. Note that the victim, offenders, community, and professional accounting bodies are all involved in the restorative process. The main goal is to reintegrate the offender back into society (Braithwaite, 2002; Beven et al., 2011; Dellaportas, 2014; Jackson, 2008). The primary victims are the investors who are most impacted by the crime. The secondary victims are the other stakeholders impacted by the crime and might include family members, friends, professional accounting bodies, and the state. Here restorative justice is seen as a process to make things as right as possible and includes a collaborative efforts by the professional accountants and their executive colleagues to attend to the needs created by the offenses such as reparation and restitutions (Dellaportas, 2014; English, 2013; Roberts, 2009). Depending on the restorative justice model used, the stakeholders affected by the crime sit in a circle to discuss the harm that has been done and arrive at an understanding of its nature, take responsibility for it, and then agree on a set of reforms to prevent recurrence, as well as to provide restitution repair the harm and reintegration back into the community (Braithwaite, 2013). Through dialogue that is intended to be receptive, a restorative approach will allow offenders to speak freely about the alleged crime, which will allow regulators to detect systemic risks. An argument can be made to the effect that the restitution arrangement made between the offenders and the victims may lapse, once the offender regains his or her grounding in the community and meaningful employment. However, the offender's escalation to being reintegrated and work as a professional accountant his dependent on whether or not he or she fulfills the restitution agreement made with the other stakeholders in the

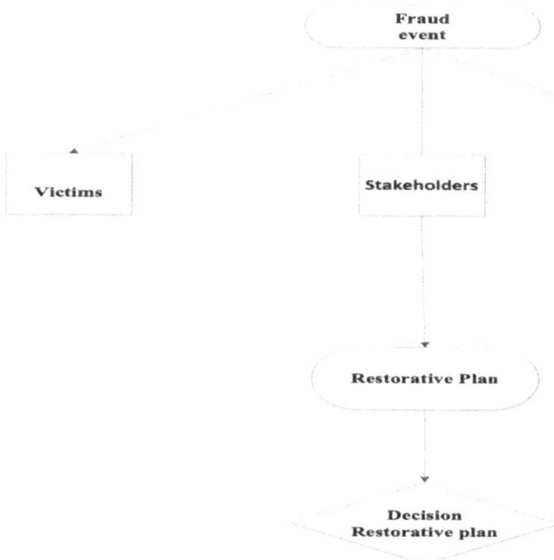

Fig. 8.1 Restorative justice framework for corporate fraud

restorative conference. To facilitate monitoring the offender, a responsive regulatory restorative approach will be used.

While this paper only permits a cursory engagement with the dialogue on responsive regulation, a suggestion is offered to demonstrate how it can be used by the state working in tandem with professional accounting associations to facilitate a process of reintegration of professional accountants (Dellaportas, 2014). The genesis of responsive regulatory theory is captured in Ayres and Braithwaite's (1992) work on responsive regulation. Ayres and Braithwaite's (1992) model focus on dialogues and collaborations that maximize the role of responsive regulation, and minimize the role of the CJS. This is not to say that the CJS's role is nonexistent; rather, it is not at the center of the enforcement process. The model they propose is called responsive regulation. It is contended that regulatory objectives can be met when enforcement agencies display a pyramid with varying degrees of interventionism. Pyramidal regulation can be one of support and one of sanction (Braithwaite, 2013). With reference to Dellaportas (2014) work, a professional accounting association that is interested in reintegrating wayward members back into the profession should first look at their strengths and seek to expand upon them.

Figure 8.2 shows the pyramid of support alongside the pyramid of sanction. Once a plan has been put in place to facilitate the reintegration and assist the professional accountants to resume their professional careers, it must be monitored for compliance. The pyramid of support helps to identify the strengths of the individual and work with the other stakeholders (i.e., professional accounting bodies and the state) to develop an action plan that will facilitate reintegration. A strength-based pyramid seeks to try one strategy after another that might further build strength on a foundation achieved on the lower end of the pyramid (Braithwaite, 2011, p. 480). The idea

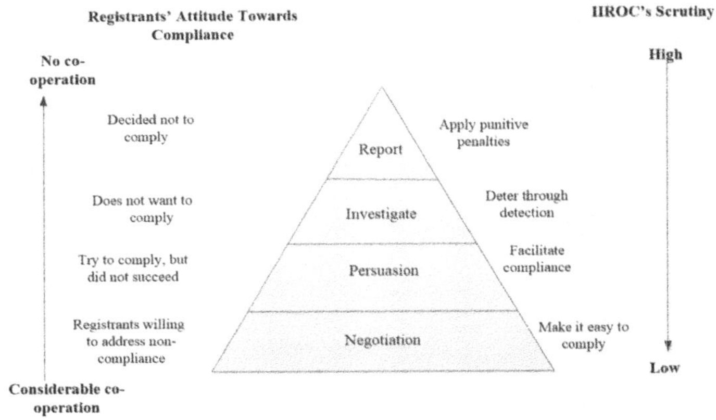

Fig. 8.2 *Source* Lokanan (2018)

of a strength-based pyramid then is that it recognizes and rewards virtue, each step representing a higher achievement than the one below and each step conferring more status than the one below. The recognition offered through a pyramid of supports to those doing the right thing plus the incentive structure provided through rewards and awards have the effect of motivating high performers to strive for excellence and average performers to do better" (Braithwaite, 2012, p. 17).

A strength-based pyramid raises the bar and provides support for the individual to get over that bar (p. 17). It nurtures hopes for the individuals to support effective reintegration and prevent future misconduct. As the individual escalate up the pyramid of support, the strengths expand until they grow to conquer risk. To put another way, we can prevent a child from falling by strapping him or her into a pram (containing risk) or we can teach him or her to walk and expand their strengths (Dukes et al., 2014, p. 289). The reintegration starts at the bottom of the pyramid of support where the offender is engage with stakeholders (in this case his or her professional accounting body) on the necessary education (i.e., ethics training) that will strengthen the quest for reintegration. The probability of being reintegrated begins when the offenders think of their self as one who would like to change by accepting responsibilities for his or her actions, acknowledge the harm done, and make reparation to the victim(s) (Dellaportas, 2014, p. 8).

8.9 Conclusion

Opportunities abound for restorative justice to inform regulatory bodies on the etiology of fraud. Regulators would secure valuable insights to monitor hot spot and prevent other corporate disaster before they materialized. This would nip in the bud the contagion of impending fraud and ferret out the deviant corporate officers and their inner circle of accountants and IT experts out of the industry. As Lokanan (2017a, 2017b) argued, corporate accounting fraud can be mitigated with controls that targets deviant behavior and tailored to the organization. Putting corporate in jail to incapacitate them is not important; putting them out of their offices, quickly, is fundamentally important. This has not happened on the scale or with the speed needed. That is standard in a crisis induced by financial oligarchs; the politics of incapacitation and cutting the losses is always painful and therefore always insufficiently ruthless with con-artists whom the politicians would still love to believe (Braithwaite, 2009, p. 443).

In the English common law, a key element is in place for alternative programs: *discretion*. A constable can legally decide not to have someone charged; a judge can refer a case to some kind of mediation. But as a practical matter, successful restorative justice programs tend to rely on some outside community organization to provide the skilled conference facilitators or mediators to make the alternative processes possible. Countries such as Canada and Australia have successfully implemented alternative models to deal with street criminality. Standard setters and regulatory bodies who are serious about reducing corporate offending can use these models to implement disciplinary system to deal with corporate offending.

So long as the regulatory officials see their goals of restorative as one to heal injuries, restored individuals and relationship, and reduce offending, then the results indicates they are on the right track. Too often, project evaluation is perceived as short-term where the immediate results of a project are paramount. If the community benefits put forward to justify the development and implementation of alternative, local justice institutions are to be taken seriously, evaluation of community projects also must become a long-term commitment. The potential for professional bodies to gain the confidence of their stakeholders and enhance relationships between the professions and the public is largely dependent on their industriousness and will to regulate and discipline offenders in the public interest.

References

Ayres, I., & Braithwaite, J. (1992). *Responsive regulation: Transcending the deregulation debate*. Oxford University Press.

Ahmed, E., & Braithwaite, V. (2004). When tax collectors become collectors for child support and student loans: Jeopardizing the revenue base? *Kyklos, 57*(3), 303–326. https://doi.org/10.1111/j.0023-5962.2004.00256.x

Ahmed, E., & Braithwaite, V. (2005). A need for emotionally intelligent policy: Linking tax evasion with higher education funding. *Legal and Criminological Psychology, 10*(2), 291–308. https://doi.org/10.1348/135532505X37263

Albanese, J. S. (1995). *White-collar crime in America*. Prentice Hall.

Ashforth, B. E., & Anand, V. (2003). The normalization of corruption in organizations. *Research in Organizational Behavior, 25*, 1–52. https://doi.org/10.1016/S0191-3085(03)25001-2

Ball, R. (2009). Market and political/regulatory perspectives on the recent accounting scandals. *Journal of Accounting Research, 47*(2), 277–323. https://doi.org/10.1111/j.1475-679X.2009.00325.x

Bazemore, G. (2005). Whom and how do we reintegrate—Finding community in restorative justice. *Criminology and Public Policy, 4*, 131.
Bazemore, G., & Mara, S. (2004). Paradigm muddle or paradigm paralysis? The wide and narrow roads to restorative justice reform (or, a little confusion may be a good thing). *Contemporary Justice Review, 7*(1), 37–57. https://doi.org/10.1080/1028258042000211996
Becker, G. S. (1968). Crime and punishment: An economic approach. *Journal of Political Economy, 76*(2), 169–217.
Benson, M. L. (1990). Emotions and adjudication: Status degradation among white-collar criminals. *Justice Quarterly, 7*(3), 515–528. https://doi.org/10.1080/07418829000090711
Beven, J. P., Hall, G., Froyland, I., Steels, B., & Goulding, D. (2011). Restoration or renovation? Evaluating restorative justice outcomes. *Psychiatry, Psychology and Law, 12*(1), 194–206. https://doi.org/10.1375/pplt.2005.12.1.194
Boyd, N. (2005). *Eron mortgage study: Final report*. https://www.bcsc.bc.ca/-/media/PWS/Resources/About_Us/Publications/Eron_Research_Study.pdf
Braithwaite, J. (1989). *Crime, shame, and reintegration*. Cambridge University Press.
Braithwaite, J. (2007). Encourage restorative justice. *Criminology and Public Policy, 6*(4), 689–696.
Braithwaite, J. (2009). Restorative justice for banks through negative licensing. *The British Journal of Criminology, 49*(4), 439–450. https://doi.org/10.1093/bjc/azp038
Braithwaite, J. (2011). The essence of responsive regulation. *UBCL Review, 44*, 475.
Braithwaite, J. (2013). Flipping markets to virtue with qui tam and restorative justice. *Accounting, Organizations and Society, 38*(6), 458–468. https://doi.org/10.1016/j.aos.2012.07.002
Braithwaite, V. (2012). Regulating for learning in Australia's Tertiary Education Institutions. Retrieved from: https://static1.squarespace.com/static/5c05f8595cfd7901fc57139d/t/5c072c6cf950b713baca1ba9/1543973999500/2012.regulating+for+learning+in+australia%27s+tertiary.pdf
Butler, B. (2012, March 22). *Courts soft on white-collar crime: Former top judge*. The Age. https://www.theage.com.au/national/courts-soft-on-white-collar-crime-former-top-judge-20120322-1vmzh.html
Calavita, K., & Pontell, H. N. (1990). "Heads i win, tails you lose": Deregulation, crime, and crisis in the savings and loan industry. *Crime and Delinquency, 36*(3), 309–341. https://doi.org/10.1177/0011128790036003002
Coleman, J. W. (1985). *The criminal elite: The sociology of white-collar crime* (3rd ed.). St. Martin's Press.
Coleman, J. W. (1987). Toward an integrated theory of white-collar crime. *American Journal of Sociology, 93*, 406–439.
Costelloe, M. T., Chiricos, T., Buriánek, J., Gertz, M., & Maier, D. (2002). The social correlates of punitiveness toward criminals: A comparison of the Czech Republic and Flori. *Justice System Journal, 23*(2), 191–220. https://doi.org/10.1080/0098261X.2002.10767665
Crawford, K. (2005, September 23). *Ebbers gets 25 years*. CNN Money. https://money.cnn.com/2005/07/13/news/newsmakers/ebbers_sentence/
Croall, H. (1992). *White collar crime*. Open University Press.
Croall, H. (2001). *Understanding white collar crime*. McGraw-Hill Education.
Crockford. (2004, January 29). *Corporations are insane*. Alternet.Org. https://www.alternet.org/2004/01/corporations_are_insane/
de Beus, K., & Rodriguez, N. (2007). Restorative justice practice: An examination of program completion and recidivism. *Journal of Criminal Justice, 35*(3), 337–347. https://doi.org/10.1016/j.jcrimjus.2007.03.009
Dellaportas, S. (2013). Conversations with inmate accountants: Motivation, opportunity and the fraud triangle. *Accounting Forum, 37*(1), 29–39. https://doi.org/10.1016/j.accfor.2012.09.003
Dellaportas, S. (2014). The effect of a custodial sentence and professional disqualification on reintegration. *Critical Perspectives on Accounting, 25*(8), 671–682.
DiMarino, F. J., & Roberson, C. (2013). *Introduction to Corporate and White-Collar Crime*. CRC Press.

Dorminey, J. W., Fleming, A. S., Kranacher, M.-J., & Riley, R. A. (2010). Beyond the fraud triangle. *The CPA Journal, 80*(7), 17–23,3.

Dukes, G., Braithwaite, J., & Moloney, J. P. (2014). *Pharmaceuticals, corporate crime and public health*. Edward Elgar Publishing.

English, L. M. (2013). The impact of an independent inspectorate on penal governance, performance and accountability: Pressure points and conflict "in the pursuit of an ideal of perfection." *Critical Perspectives on Accounting, 24*(7), 532–549. https://doi.org/10.1016/j.cpa.2013.08.002

FAIR Canada. (2011). Dishonest advisors continue to prey on investors in Canada. Retrieved on December 12, 2011 from FAIR Canada's Web site: http://faircanada.ca/top-news/dishonest-advisors-continue-to-prey-on-investors-in-canada

Findlay, M., & Hanif, N. (2012). Taking crime out of crime business. *International Journal of Law, Crime and Justice, 40*(4), 338–368. https://doi.org/10.1016/j.ijlcj.2012.05.002

Fine and punishment. (2012, July 21). *The economist*. https://www.economist.com/finance-and-economics/2012/07/21/fine-and-punishment

Fisse, B., & Braithwaite, J. (1983). *Impact of Publicity on Corporate Offenders, The*. Suny Press.

Freiberg, P. A. (2000). *Sentencing white-collar criminals*. http://citeseerx.ist.psu.edu/viewdoc/download?doi=10.1.1.507.397&rep=rep1&type=pdf

Friedrichs, D. O. (1996). Trusted criminals: White collar crime in contemporary society. *Choice Reviews Online, 34*(01), 34–0627–34–0627. https://doi.org/10.5860/CHOICE.34-0627

Friedrichs, D. (2009). *Trusted criminals: White collar crime in contemporary society*. Cengage.

Gabbay, Z. D. (2006). Exploring the limits of the restorative justice paradigm: Restorative justice and white-collar crime. *Cardozo Journal of Conflict Resolution, 8*, 421–485.

Goodstein, J., & Butterfield, K. D. (2010). Extending the horizon of business ethics: Restorative justice and the aftermath of unethical behavior. *Business Ethics Quarterly, 20*(3), 453–480.

Gottschalk, P., & Rundmo, T. (2014). Crime: The amount and disparity of sentencing—A comparison of corporate and occupational white collar criminals. *International Journal of Law, Crime and Justice, 42*(3), 175–187. https://doi.org/10.1016/j.ijlcj.2014.01.002

Grasmick, H. G., & Bursik, R. J. (1990). Conscience, significant others, and rational choice: Extending the deterrence model. *Law and Society Review, 24*(3), 837–861.

Gruner, R. (1997). *Corporate crime and sentencing* (2nd ed.). Ohio.

Hamilton, T. (2007, December 8). Solutions to market regulation faults sought. *The Toronto Star*. https://www.thestar.com/business/2007/12/08/solutions_to_market_regulation_faults_sought.html

Hansen, L. L. (2009). Corporate financial crime: Social diagnosis and treatment. *Journal of Financial Crime, 16*(1), 28–40. https://doi.org/10.1108/13590790910924948

Harris, N. (2006). Reintegrative shaming, shame, and criminal justice. *Journal of Social Issues, 62*(2), 327–346. https://doi.org/10.1111/j.1540-4560.2006.00453.x

Hartney, C. (2006). US rates of incarceration: A global perspective (fact sheet). *Research from the National Council on Crime and Delinquency*, 1–8.

Hayes, H. (2005). Assessing reoffending in restorative justice conferences. *Australian and New Zealand Journal of Criminology (australian Academic Press), 38*(1), 77–101. https://doi.org/10.1375/acri.38.1.77

Holtfreter, K., Van Slyke, S., Bratton, J., & Gertz, M. (2008). Public perceptions of white-collar crime and punishment. *Journal of Criminal Justice, 36*(1), 50–60. https://doi.org/10.1016/j.jcrimjus.2007.12.006

Investment Dealers Association of Canada v. Clarke, [2007] I.D.A.C.D. No. 7.

Ivancevich, J. M., Konopaske, R., & Gilbert, J. A. (2008). Formally shaming white-collar criminals. *Business Horizons, 51*(5), 401–410. https://doi.org/10.1016/j.bushor.2008.03.005

Jackson, H. R. (2008). *The truth in black and white*. Charisma Media.

Kappeler, V., Blumberg, M., & Potter, G. (2000). *Myths that justify crime: A look at white collar criminality prospect*. Waveland Press.

Kayes, D. C., Stirling, D., & Nielsen, T. M. (2007). Building organizational integrity. *Business Horizons, 50*(1), 61–70. https://doi.org/10.1016/j.bushor.2006.06.001

Kuo, S.-Y., Longmire, D., & Cuvelier, S. J. (2010). An empirical assessment of the process of restorative justice. *Journal of Criminal Justice, 38*(3), 318–328. https://doi.org/10.1016/j.jcrimjus.2010.03.006

Latimer, J., Dowden, C., & Muise, D. (2005). The effectiveness of restorative justice practices: A meta-analysis. *The Prison Journal, 85*(2), 127–144. https://doi.org/10.1177/0032885505276969

Lehman, C. R., & Okcabol, F. (2005). Accounting for crime. *Critical Perspectives on Accounting, 16*(5), 613–639. https://doi.org/10.1016/j.cpa.2003.08.003

Leigh, J. P., Markowitz, S., Fahs, M., & Landrigan, P. J. (2000). *Costs of occupational injuries and illnesses*. University of Michigan Press.

Levi, M. (2002). Suite justice or sweet charity? Some explorations of shaming and incapacitating business fraudsters. *Punishment & Society, 4*(2), 147–163. https://doi.org/10.1177/14624740222228527

Levi, M. (2009). Suite revenge? The shaping of folk devils and moral panics about white-collar crimes. *The British Journal of Criminology, 49*(1), 48–67. https://doi.org/10.1093/bjc/azn073

Lokanan, M. E. (2009). An open model for restorative justice: Is there room for punishment? *Contemporary Justice Review, 12*(3), 289–307. https://doi.org/10.1080/10282580903105806

Lokanan, M. E. (2014). The Investment Dealers Association of Canada's Enforcement Record: Are penalties "grossly inadequate"? *Journal of Financial Regulation and Compliance, 22*(3), 235–251.

Lokanan, M. E. (2015). Securities regulation: Opportunities exist for IIROC to regulate responsively. *Administration and Society*. https://doi.org/10.1177/0095399715584637

Lokanan, M. E. (2017a). Self-regulation and compliance enforcement practices by the Investment Dealers Association in Canada: 1984 to 2008. *Journal of Financial Regulation and Compliance, 25*(1), 2–21.

Lokanan, M. E. (2017b). Theorizing financial crimes as moral actions. *European Accounting Review*. https://doi.org/10.1080/09638180.2017.1417144

Lokanan, M. E. (2018). Securities regulation: Opportunities exist for IIROC to regulate responsively. *Administration and Society, 50*(3), 402–428.

Lynch, M. J., McGurrin, D., & Fenwick, M. (2004). Disappearing act: The representation of corporate crime research in criminological literature. *Journal of Criminal Justice, 32*(5), 389–398. https://doi.org/10.1016/j.jcrimjus.2004.06.001

Maddan, S., Hartley, R. D., Walker, J. T., & Miller, J. M. (2012). Sympathy for the devil: An exploration of federal judicial discretion in the processing of white-collar offenders. *American Journal of Criminal Justice, 37*(1), 4–18.

Makkai, T., & Braithwaite, J. (1994). Reintegrative shaming and compliance with regulatory standards. *Criminology, 32*(3), 361. http://dx.doi.org.ezproxy.royalroads.ca/https://doi.org/10.1111/j.1745-9125.1994.tb01158.x

Marceau, N., & Mongrain, S. (2011). Competition in law enforcement and capital allocation. *Journal of Urban Economics, 69*(1), 136–147. https://doi.org/10.1016/j.jue.2010.07.003

McMullan, J. L. (1992). *Beyond the limits of the law: Corporate crime and law and order*. Fernwood Publishing. https://www.ncjrs.gov/App/Publications/abstract.aspx?ID=152259

Mitchell, A., Sikka, P., & Willmott, H. (1998). Sweeping it under the carpet: The role of accountancy firms in moneylaundering. *Accounting, Organizations and Society, 23*(5–6), 589–607. https://doi.org/10.1016/S0361-3682(98)00010-5

Morris, A. (2002). Critiquing the critics: A brief response to critics of restorative justice. *The British Journal of Criminology, 42*(3), 596–615. https://doi.org/10.1093/bjc/42.3.596

Murphy, K., & Harris, N. (2007). Shaming, shame and recidivism: A test of reintegrative shaming theory in the white-collar crime context. *The British Journal of Criminology, 47*(6), 900–917. https://doi.org/10.1093/bjc/azm037

Murphy, P. R., & Dacin, M. T. (2011). Psychological pathways to fraud: Understanding and preventing fraud in organizations. *Journal of business ethics, 101*(4), 601–618.

Murphy, P. R. (2012). Attitude, Machiavellianism and the rationalization of misreporting. *Accounting, Organizations and Society, 37*(4), 242–259. https://doi.org/10.1016/j.aos.2012.04.002
Nagel, I., & Hagan, J. (1992). The sentencing of white-collar criminals in federal courts: A socio-legal exploration of disparity. *Michigan Law Review, 80*(7), 1427–1465.
Niemeyer, M., & Shichor, D. (1996). A preliminary study of a large victim/offender reconciliation program. *Federal Probation, 60*(3), 30.
Nocera, J. (2011, February 26). Biggest fish face little risk of being caught (Published 2011). *The New York Times*. https://www.nytimes.com/2011/02/26/business/economy/26nocera.html
Nugent, W. R., Umbreit, M. S., Wiinamaki, L., & Paddock, J. (2001). Participation in victim-offender mediation and reoffense: Successful replications? *Research on Social Work Practice, 11*(1), 5–23. https://doi.org/10.1177/104973150101100101
O'Connell, B. T. (2004). Enron. Con: "He that filches from me my good name ... makes me poor indeed." *Critical Perspectives on Accounting, 15*(6), 733–749. https://doi.org/10.1016/j.cpa.2004.02.002
Outterson, K. (2012). Punishing health care fraud—Is the GSK settlement sufficient? *New England Journal of Medicine, 367*(12), 1082–1085. https://doi.org/10.1056/NEJMp1209249
Parker, L. M. (2005). The smartest guys in the room: The amazing rise and scandalous fall of Enron: By Bethany McLean and Peter Elkind. Penguin Group, NY, NY, 2003. 414 pages; $26.95. ISBN 1-5918-4008-2. *Research in Accounting Regulation, 18*, 353–356. https://doi.org/10.1016/S1052-0457(05)18022-9
Perri, F. S. (2011). White-collar criminals: The 'kinder, gentler' offender? *Journal of Investigative Psychology and Offender Profiling, 8*(3), 217–241. https://doi.org/10.1002/jip.140
Piquero, N. L., Rice, S. K., & Piquero, A. R. (2008). Power, profit, and pluralism: New avenues for research on restorative justice and white-collar crime. In *Restorative justice: From theory to practice*. Emerald Group Publishing Limited.
Podgor, E. S. (2007). The challenge of white collar sentencing. *The Journal of Criminal Law and Criminology, 97*(3), 731–759.
Rebovich, D. J., & Kane, J. L. (2002). *An Eye for an Eye in the Electronic Age: Gauging Public Attitude toward White Collar Crime and Punishment, 1*(2), 19.
Revsine, L. (2002). Enron: Sad but inevitable. *Journal of Accounting and Public Policy, 21*(2), 137–145. https://doi.org/10.1016/S0278-4254(02)00044-3
Rezaee, Z. (2005). Causes, consequences, and deterence of financial statement fraud. *Critical Perspectives on Accounting, 16*(3), 277–298. https://doi.org/10.1016/S1045-2354(03)00072-8
Roberts, J. (2009). No one is perfect: The limits of transparency and an ethic for 'intelligent' accountability. *Accounting, Organizations and Society, 34*(8), 957–970. https://doi.org/10.1016/j.aos.2009.04.005
Rodriguez, N. (2007). Restorative justice at work: Examining the impact of restorative justice resolutions on Juvenile Recidivism. *Crime and Delinquency, 53*(3), 355–379. https://doi.org/10.1177/0011128705285983
Rosoff, S. M., Pontell, H. N., & Tillman, R. H. (2004). *Profit without honor: White-collar crime and the looting of America, Second Edition.*
Scott, D. W. (1989). Policing corporate collusion. *Criminology, 27*(3), 559–587.
Searcey, D., & Yuan, L. (2005, June 21). Adelphia's John Rigas Gets 15 Years. *The Wall Street Journal*. https://www.wsj.com/articles/SB111928656657864197.
Sherman , L. W. (1995). Hot spots of crime and criminal careers of place. In J. E. Eck, & D. Weisburd (eds.), *Crime and place: Crime prevention studies* , Vol. 1, pp. 35–52. Criminal Justice Press.
Sherman, L. W., Strang, H., & Woods, D. J. (2000). *Recidivism patterns in the Canberra Reintegrative Shaming Experiments (RISE)*. Research School of Social Sciences, Australian National University.
Sikka, P. (2009). Financial crisis and the silence of the auditors. *Accounting, Organizations and Society, 34*(6), 868–873. https://doi.org/10.1016/j.aos.2009.01.004

Simpson, S. S. (2002). *Corporate crime, law, and social control*. Cambridge University Press.
Simpson, S. S., Gibbs, C., Rorie, M., Slocum, L. A., Cohen, M. A., & Vandenbergh, M. (2013). An empirical assessment of corporate environmental crime-control strategies. *The Journal of Criminal Law and Criminology (1973–), 103*(1), 231–278.
Strader, J. K. (2007). White collar crime and punishment: Reflections on Michael, Martha, and Milberg Weiss. *George Mason Law Review, 15*, 45.
Thomas, K., & Schmidt, M. S. (2012, July 2). Glaxo agrees to pay $3 billion in fraud settlement (Published 2012). *The New York Times*. https://www.nytimes.com/2012/07/03/business/glaxosmithkline-agrees-to-pay-3-billion-in-fraud-settlement.html
Trotman, A. (2014, January 7). *JP Morgan raised concerns over Bernard Madoff 10 years before he was arrested*. The Telegraph. https://www.telegraph.co.uk/finance/financetopics/bernard-madoff/10556868/JP-Morgan-raised-concerns-over-Bernard-Madoff-10-years-before-he-was-arrested.html
Umbreit, M. S., & Vos, B. (2000). Homicide survivors meet the offender prior to execution: Restorative justice through dialogue. *Homicide Studies, 4*(1), 63–87. https://doi.org/10.1177/1088767900004001004
United States Sentencing Commission, (2004). *United States Sentencing Guidelines 2004*. https://www.ussc.gov/sites/default/files/pdf/guidelines-manual/2004/manual/gl2004.pdf
Weisburd, D., Waring, E., & Chayet, E. (2001). *White-collar crime and criminal careers*. Cambridge University Press.
Weissmann, A., & Block, J. A. (2007, February 21). *White-collar defendants and white-collar crimes*. The Yale Law Journal. https://www.yalelawjournal.org/forum/white-collar-defendants-and-white-collar-crimes
Why have so few bankers gone to jail? (2013, May 13). *The Economist*. https://www.economist.com/the-economist-explains/2013/05/13/why-have-so-few-bankers-gone-to-jail
Wolfe, S. M. (2013). Escalating criminal and civil violations: Pharma has corporate integrity? Not really. *British Medical Journal, 347*, f7507. https://doi.org/10.1136/bmj.f7507
Wright, V. (2010). Deterrence in criminal justice: Evaluating certainty versus severity of punishment. *The Sentencing Project*, 1–9.
Zehr, H. (2015). Changing lenses: *Restorative justice for our times*. MennoMedia, Inc.

Part IV
Recent Issues on Boards, Gender Issues

Chapter 9
Female Entrepreneurship: Is It Possible to Reconcile Family and Business?

Karine Rosália Felix Praça Gomes and Emanuel Ferreira Leite

Abstract The desire for professional achievement and independence, the unemployment of a partner or the lack of increased family income have brought women into the business world. The objective of the research is to conduct a study on female entrepreneurship, considering the obstacles faced by women who are committed to harmonizing business and family. The research used was qualitative through literature review and case study. The instruments applied were classroom observation and the questionnaire applied to a certain population who participated in the III Entrepreneurship Fair held at the Faculty of Science and Technology, Professor Dirson Maciel de Barros (FADIMAB) in Goiana-PE. According to the selected sample, more than 41.67% of respondents are women and 58.33% are men. With the work it was possible to infer that of the various attributes that entrepreneurial students have in common, there are certain characteristics such as self-realization, persistence, ability to create, positivity, entrepreneurial vision and courage to achieve their goals and overcome obstacles. The data obtained confirms the literature on female entrepreneurship, that the female public innovates to face obstacles, introducing changes in the family, professional and personal spheres. The study presents important contributions to the analysis and understanding of the phenomenon of entrepreneurial behaviour and female entrepreneurship in the local context analysed, shedding light on a reality still little explored from an academic and scientific point of view.

Keywords Higher education · Feminine intention · Plurality of tasks · Family work relationship

K. R. F. P. Gomes (✉)
Federal Rural University of Pernambuco, Pernambuco, Brazil
e-mail: karine.rosalia@ufrpe.br; karine.rosalia@upe.br

E. F. Leite
University of Pernambuco, Pernambuco, Brazil

M. A. Naser (ed.), *New Approaches to CSR, Sustainability and Accountability, Volume IV*, Accounting, Finance, Sustainability, Governance & Fraud: Theory and Application, https://doi.org/10.1007/978-981-16-9499-8_9

9.1 Introduction

Research on gender and entrepreneurial actions has stimulated academic curiosity, developers of public intervention actions and development structures in recent years (Feder & Nitu-Antonie, 2017; Thébaud, 2015). Demands such as gender inequality in entrepreneurship signal one of the relevant purposes in the literature.

In the economic and administrative sciences, there is a broad concept theory, also called entrepreneurship. The concepts are specifically disposed in the generation of new entities, in the survey of favourable occasions, innovation, creativity and production of goods and services (Druker, 1974; Venkataraman, 2019). Entrepreneurship has an important position not only in the origin of new ventures, but also in progress and local growth. Therefore, entrepreneurship requires a conduct that is based on a conception that succeeds in the idealization of innovative products/processes or the entry into a new business. In this sense, the subject who identifies a given scenario as a favourable occasion, makes its evaluation and exploitation, is considered an entrepreneur.

In higher education, it is observed that interventions of conventional and unconventional education instigate the growth of the desire to undertake among students (Casero et al., 2017). However, Sousa et al. (2016) argue that "not only changes in professional occupation or employment have highlighted the last few years, but also changes in the profile of the intellectual patrimony made available, regarding the search for qualification in higher education courses" (p. 91). Such a scenario explains the progressive access to higher education institutions, including those already included in the labour market.

Higher education institutions and centres of higher education with entrepreneurial functions are considered favoured places for the propagation of these perspectives, in their instructional procedures and investigations (Glavic & Lukman, 2007). The university space by "providing entrepreneurial behaviour will collaborate decisively in the disposition of an entrepreneurial culture, driving the forces towards the valorization and encouragement of those who intend to undertake" Casado et al. (2012). Thus, it is reasonable to assimilate that the higher education environment has a significant attribution in the understanding of the relationships between teaching entrepreneurship and starting a business (Bae et al., 2014; Martin et al., 2013). In the perception of these authors, this is a way to achieve sustainable development through entrepreneurial education and citizen action by enabling, in a way that can include, equally and safely, economic development, social promotion and environmental preservation.

The objective of the research is to conduct a study on female entrepreneurship considering the obstacles faced by women who undertake to harmonize a business and the family. The educational practices and experiments directed towards the creation of the enterprise itself in higher education scenarios are capable of being effective in promoting entrepreneurship, if the distinctions and peculiarities between the different areas are taken into consideration (Borges, 2014). In this sense, the importance of this research is based on the concept that students see in the university space a range of possibilities that allow them to apply their abilities and aptitudes.

The organization of the research is a brief introduction. Soon after, the theoretical basis that addresses points related to the concepts leading this study. Next, the methodological procedures used are clarified. Afterwards, the empirical research is examined, the results obtained and the conclusion about the study are presented. Finally, the references used are indicated.

9.2 Methodology

The qualitative research is characterized by means such as a case study, which has a unit of study, to conduct an examination of the business profile of students during Business Administration at the Faculty of Science and Technology, Professor Dirson Maciel de Barro [FADIMAB].

The instruments used were classroom observation and a questionnaire specially prepared for this research. Such instruments seek to cover the subject in question, while offering ways of building an idea by relating situations that are relevant to the personal experience of the sample participants. The questions were structured in the Likert option, presented on a five-point scale, covering the frequency of: 1—Unimportant; 2—Unimportant; 3—Medium important; 4—Very important; 5—Extremely important.

To this end, 150 students enrolled in the Administration course of FADIMAB understood the universe of research, but to verify the study, a sample of 60 students was analysed, of these only 25 participants, because they agreed with the proposal of the study. This sample was constituted by a non-probabilistic process, based on the acceptance of the subjects to participate in the research. Therefore, the number of questionnaires was not pre-defined, and the closing of the collection of information was based on the judgement of the researchers regarding the redundancy of the problems raised.

The instrument was made collectively and personally during the event of the 3rd FADIMAB Entrepreneurship Fair, an event offered as part of the pedagogical week. The students answered the questionnaire after explaining the objectives of the research. Regarding the analytical categories, depending on the objectives of the research, the categories and elements of analysis were defined and presented in Table 9.1.

Descriptive statistical analysis tools were used to examine the general profile of entrepreneurs and characteristics of the business, as well as the profile and behaviour of the women investigated. These techniques will be used to achieve the objectives of the research, especially about the delimitation of a scenario, even if initial, the manifestation of female entrepreneurship in the city of Goiana-PE, Brazil.

Table 9.1 Analytical categories and elements of analysis

Categories	Elements of analysis
Profile of the entrepreneur	Age; education; marital status; number of children; previous experience; characteristic of the company
Behaviour of women entrepreneurs	Opportunities and initiative, commitment, demand for quality and efficiency, systematic planning and monitoring, calculated risk taking, information seeking, persuasion and networking, goal setting, persistence, independence and self-confidence and the challenges faced
Conflicts between work and family	Time dedicated to the family, the company and herself; perception of the efficiency of the roles performed; greater conflicts existing at this time; strategies adopted to manage conflicts; relationship with spouse and children; presence of the couple. (a) as a minimizing effect on the conflict

Source Authors' elaboration (2019)

9.3 Discussion and Results

9.3.1 Entrepreneurship in the Higher Education Sphere

The concept of entrepreneurship in the academic environment indicates that both its origin and its progressive development are permeated by historical particularities of a social, economic and political nature, in addition to being highlighted by the many disciplinary visions that are implicit in them. Joseph Schumpeter (1961), who emphatically inspired the scientific reflection around the theme, gave the entrepreneur the mission of sustaining the capitalist framework through creative destruction, promoting, consequently, its relevance in terms of the vital force for the economy. The creative destruction would be the result of the respective evolution of the economic progression and its stages and would result mainly from innovation. In Zanchet's (2019) conception, entrepreneurship is one of the principles capable of assisting a nation's economic, social, technological and educational impulse. For authors Behling and Lenzi (2019), entrepreneurship is one of the principles promoting socioeconomic growth, by increasing the dynamism of the economy and cooperating for social welfare, through innovation in products and services that supply people's demands.

For Drucker (1974) entrepreneurship is: practice, market vision, evolution, and states that the:

> "...the specific work of entrepreneurship in a business enterprise is to make today's businesses capable of doing the future, becoming a different business" [...]Entrepreneurship is neither science nor art. It is a practice. (Druker, 1974, p. 25)

Following this same meaning, Leite (2017) corroborates when it states that being an entrepreneur means having the aptitude to propose or realize something, fertile imaginative capacity to create ideas, versatility to adapt them, creative capacity to convert them into a business opportunity, instigation to think in a conceptual way, and the aptitude to see, capturing with intelligence the change as a favourable occasion, taking seriously the management of your business.

In this sense, whatever the action in entrepreneurship in the higher education institution needs to be systematized aiming at the empowerment of the student and the ratification of his/her main role in the university. In this way, the training in higher education needs to be transversal, so that the students conclude their graduation having all the indispensable instruments for a commitment and proactivity conduct and being able to have a better condition to be employed in the market.

9.3.2 Women's Intention to Undertake

Research on gender and entrepreneurial action has attracted the attention of higher education communities, agents who formulate public policies, and development agencies in recent years. According to Gómez Gómez Granada et al. (2019), young people translate entrepreneurship as procedures to be, do, have, know and still confront it as a vital purpose. In view of this, entrepreneurship is an action that provides and conceives processing of man's progress, strives to improve satisfaction and ensures quality of life.

The competence of creating and managing a grouping of goods is no longer the exclusive merit of men, the female public with a systemic look and an extremely intuitive feeling has been putting into practice in the most varied parts of the market, taking care of strategic functions and contracting accepted among their equals.

The strength of the recent paths in the feminine trajectory comes from a gathering of aspects that cross preliminary of the effort to overcome their independence, accompanied by the degree of education that has made possible greater social and economic autonomy. In another follow-up to the progressive development of women's activities, entrepreneurship emerges as a support to free women from traditional work for transformation and institution of enterprises that aim to balance work needs according to family demands.

The debates in the scientific field around the function and performance of women in the labour market, especially in innovative enterprises, show their relevant performance in the economic development and in the benefit of the well-being of the families supported by these women (Machado et al., 2016; Nassif et al., 2014), which support several threats to adjust the profession to the particular objectives.

It is also worth mentioning the contribution made by Gomes et al. (2011), who sought to understand how women identify the circumstances that led them to entrepreneurship, establishing as fundamental reasons, personal promotion, the intuition of a favourable business opportunity, the obstacle of promotion in the profession, the lack of sustenance and finally the viability of combining work and family, this being the final outline that should be highlighted.

Feminine entrepreneurship is expanded due to a series of possibilities and demands perceived by these women, either in the way of representation as protagonist or collaborator, it is explicit their competence to identify favourable occasions and undertake ideas in the format of recent business having the opportunity to unite their personal achievement and professional-financial emancipation.

It is evident that entrepreneurship plays an emancipatory role in the existence of women, who when creating their own enterprise experience a mixture of feelings of satisfaction with the probabilities of facing new obstacles related to the execution of their particular and professional demands.

9.3.3 Female Entrepreneurship and the Family-Work Relationship

In our country, the most current studies treat the subject of "female entrepreneurship" in a way that contributes to stereotypes when they are limited to making analogies between the genders (Gomes et al., 2009). However, the female public has gradually gained a place in countless sectors, sustaining that woman have been occupying spaces of evidence at the social, economic, cultural and political levels.

In the sphere of entrepreneurship, however, this evolution is reflected in what Gomes et al. (2009) point out that one of the predominant reasons why women come to own the enterprise itself is the malleability of schedules, because in this way they will be able to make family and work compatible.

Strobino and Teixeira (2014), on the other hand, find it difficult to find women who undertake the limit between personal life and work, or family life, well defined, and as a result, in general, the family-work conflict is confronted.

When discussing work-family reconciliation, Oliveira et al. (2013) look at models conceived in the mind that draw a parallel between an idealized observation and a theory: the work-family conflict that happens when work obligations affect family life, and the family-work conflict that happens when it is the family issue that causes harm to continued work.

These authors share the opinion that understanding this relationship is important and can contribute to institutions providing appropriate means for the harmony of these two environments. That said, there can be no disagreement that in the institutional space it is essential for the good institutional functioning that the participant, woman or man, is able to manage private and professional life in such a way that they do not disharmonize. Therefore, it is examined that, in the management of this conflict, female entrepreneurs are in inequality of advantages when compared to men.

According to Gomes et al. (2011), the demand to reconcile their multiple roles with the sacrifice of professional advancement has been encouraging women to effectively perform the entrepreneurial capacity for themselves. Corroborating the theme, Shelton (2006) considers that the softening or eradication of the family-labour conflict is a relevant requirement in promoting the development of the enterprise. Therefore, the author presents three main variables of family-work management.

The number one is the eradication of the conflict that matters in not establishing a family or business; the number two strategy is the softening of the conflict, or better, forming a smaller family and having third parties as support in the fulfilment of the tasks; finally, the number three is the distribution of the conflict, which implies in transferring tasks to the family members or their employees.

In their studies, Strobino and Teixeira (2014) expose the work-family conflicts identified by two entrepreneurs in the construction material business in the city of Curitiba. Considering the variables: time, tension and behaviour of the theoretical form of Greenhaus and Beutell (1985) and the procedures to soften the conflict presented by Hall (1972) and Shelton (2006). The authors made use of the technique of diversified situations in the study with qualitative treatment and data collection through semi-structured interview. With the information collected, the study proved that the variable time is the most frequently cited as the cause of work-family conflicts and emotional control as the most mentioned expression as moderator of these respective conflicts.

9.4 Analysis of Results

It can be presented as a response to the data collected regarding the profile of entrepreneurs, that the interviewed students had a good level of education attending undergraduate courses, continuous training in the field of applied social sciences, more precisely in the field of Administration, and 20% of the participants were already in their second degree.

The sample fragment was composed of young women, on average 25 years old, of which 24% were married, 72% were single and 4% were separated. According to the marital status and age of the student businessmen, 84% had children (one child on average), confirming that 64% were between 5 and 11 years old, 12% were between 1 and 4 years old and only 20% were over 12 years old. Practically all participants received help with domestic chores (64% were supported by family members and 20% received help from others).

Regarding the businesses analysed, the data show that 20% of the companies represent the service business (beauty, buffet, insurance, English courses), 76% of the companies are engaged in the business (food, beverages, flavours), trousseau items, clothing, gymnastics, custom). The companies are led by 2 partners on average and tended to establish themselves well in the market, which shows that they have been in the position for over 3 years. As a rule, they were small companies, with about 2 employees in most of them.

When comparing the motivations that influenced the opening of the company, it was observed that some characteristic peculiarities in the studies with focus on gender were present, such as time flexibility and family reasons, but these were not the only ones. Besides these, others were present, such as: encouragement to friends, positive user experience in the business sector, desire to change the line of business, self-confidence, influence of business friends, dreams, adventurous spirit, corporate educational training.

As for the category of entrepreneurial behaviour, all participants interviewed considered that the development of autonomy and trust behaviours is extremely important, representing 80% of responses. Even in relation to the behaviours considered extremely important, goals are established, and information is sought with 48% of the answers, a favourable opportunity is sought and strategies to achieve convincing power with 60% of the answers. Also considered extremely important were 64% of the answers regarding the questions of periodic review of the business plan, personal sacrifice to perform tasks and the requirement of knowledge on how to provide their products, which allows academics to develop autonomy, have a short- and long-term vision, pursue objectives and create measurable goals.

It is also worth mentioning that 76% of participants identified quality and efficiency as extremely important behaviour. Therefore, as entrepreneurs, they are always willing to improve their business or their products.

As for the analysis on time management initially, the results obtained from the elements of analysis: time dedicated to the company, to the family and to oneself. It was found that the student businessmen had a high level of participation, a factor highlighted by their extremely dedicated participation of about 72% of their business.

Other data also revealed, in relation to the question of time dedicated to the company, that 68% of women would love to maintain the time they already dedicate to work, while 20% would enjoy spending less time and 12% intend to expand their daily effort in the company. Therefore, for a part of the women in business, the business hours seem to be in the opposite direction of desire and can be characterized as a principle of dissatisfaction.

Managing all these conflicts is not an easy task, but employers create strategies to manage conflicts and address the balance between work and personal life.

9.5 Conclusion

The performance of the female public in Brazilian entrepreneurship is quite considerable and requires a greater perception of the difficulties faced by the female public they undertake and the effects of these on their well-being.

This research indicates high involvement of women with their business. They show to be daring and determined, although apprehensive with factors related to finances, with the progress of the business, such as the contentment of consumers and the economic and socio-political conjuncture of Brazil. Their fears and responsibilities tend to be weakened by an intense feeling of victory. Envious of their accomplishments, the female public they undertake feels like a winner, either by the personal confirmation that they are the object, or by the success obtained in the consolidation of their various functions.

It is undeniable that the female gender has been changing its social actions and inventing novelties and innovative ways to face the diversity of roles, which demands its entire performance in command of the home. We proved that the reality experienced by the entrepreneurial students is not different from those experienced by ordinary women who have chosen to open a business. They work with the difficulties of any business are still women who also need to face the machismo entrenched in the social body that still sees in the female gender a being that depends on the "strength" of the man and that have no capacity to treat the performance in this space.

The profile detected shows that these entrepreneurs are fascinated by what they accomplish and are qualified to work in this space had with masculine, overcoming machismo and conquering success adjusting family and business. The data obtained ratifies the literature about female entrepreneurship, which innovate in facing obstacles, inserting changes in the family, professional and personal spheres.

Feminine action in entrepreneurship is very relevant and needs a considerable understanding of the demands faced by entrepreneurs and the consequences of these on their quality of life. Business, children and self-respect are—practically the same—causing high indicators of well-being, pointing out that the universe of profession, family and private cause timely influence on women's well-being.

Considering the peculiarity of the feminine space, it was evidenced that the prejudice of sex is a source of discomfort that affects some entrepreneurs. The activities of the multiple functions prove to be a point surrounded by a certain ambiguity, since sometimes it is related to a sensation of victory and fulfilment, others to a sensation of failure or affliction. Nevertheless, the assertive behaviour and self-assertion of the simultaneous importance of diversified aspects of life space, perceived in the women they undertake, make us assume that the performance of entrepreneurship, based on independence and decision power, puts women in a situation able to modify adversities and dissatisfactions in possible stimuli to produce satisfaction.

It is important to note that the harmony between the collective and the particular does not need to be the responsibility to be in charge of women alone. With the intention of balancing the work-life relationship, new forms are required, which provide better indicators of contentment as well as for the male gender as for the female. Thus, society must once again align the universe of professional occupation with the real life of the modern family.

Finally, it is valid to emphasize that the present research has some limiting points. In principle, by limiting the gathering of information only with the female public of the city, one has a look focused on female entrepreneurship in the scenario of the sample space investigated, but without likelihood of further perception of this reality.

In this context, future investigations may use the same theoretical-methodological instrument, adding a sample made up of both male and female gender, to assess similarities and dissimilarities as well as the entrepreneurial conduct and gender.

Moreover, the sample size, constituted through a non-probabilistic procedure by agreement of the participants, was not able to extrapolate the findings to the local population, not allowing the application of more elaborate statistical techniques. In view of this, future investigations such as examining the entrepreneurship of gender, looking for what are the peculiarities of the male and female entrepreneurs, investigate the female entrepreneurship of formal and informal entities, study the reasons for success in entrepreneurship by observing the profile and conduct of these women.

However, even with these limiting factors, the research brings relevant contributions to the investigation and understanding of the occurrence of the entrepreneurial conduct and the female entrepreneurism on the local scenery observed, clarifying a reality until now little examined on the academic and scientific approach.

References

Bae, T. J., Qian, S., Miao, C., & Fiet, J. O. (2014). The relationship between business education and business intentions: The meta-analytic review. *Theory and Practice of Entrepreneurship, 38*(2), 217–254.

Behling, G., & Lenzi, F. C. (2019). Entrepreneurship skills and strategic behavior: A study with microentrepreneurs in an emerging country. *BBR. Brazilian Business Review, 16*(3), 255–272.

Borges, L. O. (2014). Should we be against or in favor of junior companies in psychology? *Revista Psicologia Organizações e Trabalho, 14*(4), 481–488.

Casado, F. L., Siluk, J. C. M., & Zampieri, N. L. V. (2012). Entrepreneurial university and sustainable regional development: proposal of a model. *Revista de Administração da UFSM, 5*(special edition), 633–650

Casero, J. C. D., Portillo, A. F., Escobedo, M. C. S., & Mogollón, R. H. (2017). The influence of University context on entrepreneurial intentions. *Entrepreneurial Universities,* 65–81.

Drucker, P. F. (1974). *The effective manager*. Zahar Publishing House.

Feder, E., & Nitu-Antonie, R. D. (2017). Connecting gender identity, business training, role models and intentions. *International Journal of Gender and Entrepreneurship, 9*(1), 87–108.

Glavič, P., & Lukman, R. (2007). Review of sustainability terms and their definitions. *Journal of Cleaner Production, 15*(18), 1875–1885.

Gomes, D. T; Guerra, P. V., & Vieira, B. N. (2011). The challenge of feminine entrepreneurship. In *Annals of the ANPAD meeting*, p. 35. Rio de Janeiro.

Gomes, A. F., Santana, W. G. P., Araújo, U. P., & Martins, C. M. F. (2014). Feminine entrepreneurship as a research subject. *Revista Brasileira De Gestão De Negócios, 16*(51), 319–342.

Gómez Gómez Granada, B. M., Soler, J. L. B., & Zapata, B. E. Á. (2019). Learning and human development: A look from the youth. *Aletheia. Revista de Desarrollo Humano, Educativo y Social Contemporáneo, 11*(2), 127–152.

Greenhaus, J. H., & Beutell, N. J. (1985). Sources of conflict between work and family roles. *Academy of Management Review, 10*(1), 76–88.

Hall, D. T. (1972). A model of coping with role conflict: The role behavior of college educated. *Administrative Science Quarterly, 17*(4), 471–486.

Leite, E. (2017). *The phenomenon of entrepreneurship.* São Paulo: Saraiva. Available at: https://books.google.com.br/books?hl=ptBR&lr=&id=ZStrDwAAQBAJ&oi=fnd&pg=PT6&dq=empreendedorismo&ots=dJ1OHlKvpM&sig=8V7cKyomw5oKtHNo2nxN5KUeRjo#v=onepage&q=empreendedorismo&f=false> Access at: 24/02/2019.

Machado, H. P. V., Faia, V. S., & Silva, J. D. (2016). Alerta Empreendedor: Estudo da Influência de Características do Indivíduo e do Empreendimento. *Brazilian Business Review, 13*(5), 87.

Martin, B. C., McNally, J. J., & Kay, M. J. (2013). Examining the formation of human capital in entrepreneurship: A meta-analysis of entrepreneurship education outcomes. *Journal of Business Venturing, 28*(2), 211–224.

Nassif, V. M. J., Andreassi, T., & Tonelli, M. (2014). *Women entrepreneurs in situations of extreme difficulty: A study that uses the technique of critical incidents*. Organizational Studies Meeting ANPAD-EnEO, Gramado, RS.

Oliveira, L. B, Cavazotte, F. S. C. N., & Paciello, R. R. (2013). Background and consequences of labor and family conflicts. *Revista de Administração Contemporânea. Rio de Janeiro, 17*(4), 418–437.

Schumpeter, J. A. (1961). *Capitalism, socialism and democracy*. Harper and Brothers.

Shelton, L. M. (2006). Women entrepreneurs, work-family conflicts and business performance: New ideas about the work-family interface. *Small Business Management Magazine, 44*(2), 285–297.

Souza, G. H. S. D., Santos, P. D. C. F. D., Lima, N. C., Cruz, N. J. T. D., & Lezana, Á. G. R. (2016). Entrepreneurial potential and business success: A study on elements of convergence and explanation. *RAM. Mackenzie Management Magazine, 17*(5), 188–215.

Strobino, M. R. C., & Teixeira, R. M. (2014). Feminine entrepreneurship and the work-family conflict: A study of multiculturalism in the civil construction sector of the City of Curitiba. *Revista Administração, USP, São Paulo, 49*(1), 1–18.

Thébaud, S. (2015). Business as plan B: Institutional foundations of gender inequality in entrepreneurship across 24 industrialized countries. *Administrative Science Quarterly, 60*(4), 671–711.

Venkataraman, S. (2019). *The distinctive domain of research in entrepreneurship*. Emerald Publishing Limited.

Zanchet, R. E. R. (2019). The teaching of entrepreneurship in the undergraduate courses of the Federal University of the Greater Gold: a case study.

Chapter 10
Women on the Board of Indian It Companies: Are They Audible and Visible?

Fincy Pallissery, Chaitra Rangappa Beerannavar, and Freddy Thomas

Abstract Gender disparity on the board of Indian IT companies is a continuing saga despite the Indian Companies Act, 2013 mandating at least one-woman director in the executive position of public listed companies. Women in India are leaders of varied sectors of businesses and on top leadership positions except in IT companies. This study has found that 14 out of 25 leading IT companies in India has, got not more than two women on the board of directors. It has also been found that the reasons for said nomination are due to the statutory compulsion to have women on board. Information technology companies are responsible for innovation, business growth, transformation and diversification. Strategic leadership is the key for IT companies to achieve the above-stated objectives. Inclusiveness in the economic reforms is possible when women are given adequate representation in entrepreneurship and leadership positions in all sectors of industries. This study aims at examining the causes of the inadequate representation of women in Indian IT companies. Paper has examined the following issues to analyse the above-stated proposition: (a) How far the Indian IT industry has contributed to the empowerment of Indian women? (b) Whether the employment terms and recruitment policies of IT companies are sufficient to ensure the security of tenure and promotion to women employees and do it incentives women employees' contribution towards innovation in their respective companies? (c) What are the factors contributing to women taking up leadership positions in Non-IT Industries? (d) Whether family commitments are the reasons for women in the IT sector to decline leadership positions or whether male domination is a cause for women to be backward in IT companies' leadership positions? (e) Should mandatory reservation of adequate percentage of seats for women in administration be uniformly applied also for employment of women in the IT sector and how far

F. Pallissery (✉) · C. R. Beerannavar · F. Thomas
School of Law, CHRIST (Deemed to Be University), Bangalore, India
e-mail: fincy.vinoj@christuniversity.in

C. R. Beerannavar
e-mail: chaitra.beerannavar@christuniversity.in

F. Thomas
e-mail: freddy.thomas@christuniversity.in

M. A. Naser (ed.), *New Approaches to CSR, Sustainability and Accountability, Volume IV*, Accounting, Finance, Sustainability, Governance & Fraud: Theory and Application, https://doi.org/10.1007/978-981-16-9499-8_10

the practice followed in developed jurisdictions need to be incorporated under the Indian law? Enrolment of women in IT and business education is on par with their counterparts. Since women occupying leadership positions is negligible, the paper examines the challenges and proposes solutions to ensure gender equitable reforms in the leadership roles of Indian IT companies. Data related to board composition and shareholding patterns of Indian IT companies are looked into and analysed to identify whether women possess capital or management control in the Indian IT companies. To critique the role of women in other sectors of employment with that of the IT companies, data collected from the National Association of Software and Services Companies (NASSCOM) and Indian government-sponsored schemes are considered. The data are also collected from various sources such as Sustainability Reports of Wipro, Infosys, HCL, Dell, Accenture, Tata, Human Development Index (HDI), United Nations Department of Economic and Social Affairs (UNDESA). The data compiled reflect the factors that affect women's career progression in the Indian IT sector. This study has found that there is an absolute imbalance in terms of gender diversity on the boards of Indian IT companies. Reasons for the same are as follows: 1. Women who have excelled in technological education are not willing to take up leadership positions in IT companies due to the challenges and risks involved in this specific sector, 2. Family commitments and health issues are not conducive for women to dedicate the required time in managing corporate boards of IT companies, 3. Joint families and a patriarchal Indian system limits woman to undertake employment, 4. Women with liberal outlook and merit are not preferred as a choice by male leaders of IT companies due to the fact that they never want to be led by women, 5. The upskilling programmes organized by IT companies to their women employees are not sufficiently focused to promote women to leadership positions and 6. Excess share qualification for directorship prescribed by listed public companies is an impediment for women to be considered for executive positions. Paper suggests strategies and policies for the promotion of women employees to executive positions and ensuring the disclosure of diversity of corporate boards as a prerequisite to listing its shares. Secondly, it proposes to amend Companies Act, 2013 to prescribe a higher number of mandatory appointments of women on board to make it mandatory for women to be part of committees of the board mandated under the Companies Act. Thirdly, it proposes that the B-Schools admission policies should increase the intake of women candidates for management programmes so that they would possess the adequate competency to govern corporate boards of Indian IT companies.

10.1 Introduction

Women as Harbingers of Major Social Change: "Agency" and "Well-being" are the two salient features of human life. The agency aspect relates to the quest of goals and objectives that a person has reason to value and advance, irrespective of their connections to his or her own well-being. For many years, the focus of feminist literature and various women's movements was "well-being" or "protection" of women, given the "unfair deal" they have been, and are in many ways to this day, meted out. But, in due course, the objectives of these movements have gradually shifted from a narrow "welfarist" focus to a broader outlook stressing on "the active role of women as agents in doing things, assessing priorities, scrutinizing values, formulating policies and carrying our programmes".[1] It is obvious that the active agency role of women calls for the urgency of rectifying many sociopolitical influences that blight the well-being of women and subject them to deprivations of various kinds. It is clear from available empirical evidence that the biological and non-biological consequences of neglecting women's interests extend far beyond the well-being of women only, and the traditional economic approaches, viz. resource-based and preference-based, were found to be not only ineffective but to be frequently reinforcing the inequalities that are highly relevant to women's well-being. This led to the "capabilities approach"-an approach to quality-of-life assessment pioneered within economics by Amartya Sen and popularised by the highly influential Human Development Reports of the UNDP.[2] The central question addressed by the capabilities approach is not "How satisfied is this woman?" or even "How much in the way of resources is she able to command?". It is instead, "what is she actually able to do and to be?" Users of this approach ask: "Is the person capable of this, or not?" They ask not only about the person's satisfaction with what she does, but about what she does, and what she in a position to do (what her opportunities and liberties are). They ask not just about the resources that are present, but about how those do or do not go to work, enabling the woman to function.[3]

The expansion of women's capabilities not only enhances women's own freedom and well-being, but also has many other effects on the lives of all.[4] Active agency of women can, in many circumstances, lead to a substantial improvement in the loves of all people-men as well as women, children as well as adults. There is also new evidence that functioning of women in other areas, including in economic and political fields, makes a radical difference to the social outcome. Substantial links between women's agency and social achievements have been noted in many different countries, and there is plenty of evidence that whenever the social and economic arrangements move away from the standard "male-driven" models, women can seize business and economic initiative with much success. It is also, by now, clear that the result of women's participation is not merely to generate income for women, but also to provide many other social benefits that come from their enhanced status, enterprise, and independence. The success stories of Grameen Bank and BRAC in Bangladesh, originally conceived to improve the social and economic roles of women, speaks

---

[1] Sen (2006).

[2] UNDP. Various years. *Human Development Report 1993–1999*. New York, NY, United Nations Development Programme.

[3] Nussbaum (1999).

[4] Nussbaum (2000).

volumes of not just about how women's agency can help to transform the lives of all human beings but an entire nation.[5]

10.2 Women Entrepreneurship to Leadership—Historical Evidence from India

Marquis de Condorcet, a French mathematician, and a leading intellectual leader of the French revolution, and one of the pioneers who initiated the formal discipline of social choice theory,[6] was one of the first passionate advocates of public education and particularly women's education, in the 1780s. His insistence on the importance of women's education was linked, among other things, to his recognition of the need for women's voices in public affairs as well as in family and social life. This, he believed, will lead to giving priority in public policy to women's education as part of the promotion of justice in society, which will, in turn, result in direct benefits and indirect positive consequences for the general welfare of the society. The role of women's education and enlightenment was thus central to Condorcet's approach[7] to society of his times, and he even saw that a more widespread women's education would reduce the population growth rate dramatically and could even halt or reverse it, as against his more popular successor Robert Malthus[8] and the alarmist theory of population catastrophe.

However, Indian ancient recorded history has a different story to tell and it abounds of examples of women who have scaled the heights of loquacity, valour, power and influence. Even though India has had a history of deep inequalities along the lines of gender, class, caste and community, participation of women in both political leadership and intellectual pursuits is by no means negligible, notwithstanding the fact that men have tended, by and large, to rule the roost in socio-economic and political matters. Vocal leadership of women can be traced back to the Upanishads,[9] for example, in the *Brihadaranyaka Upanishad* where we are told about the famous "arguing combat" where *Yajnavalkaya*, the outstanding scholar and teacher has had to face questions of the sharpest intellectual nature, from a woman scholar *Gargi*, and more recently the Rani of Jhansi—the feminine hero—who fought along with the mutineers in the middle of the nineteenth century against the British rule.

[5] In India, however, there were reports that the for-profit banks hijacked the "microcredit" scheme as they saw there "was money at the bottom of the pyramid", until it was stopped by a spate of farmer suicides. See Prahlad (2005).

[6] Social Choice theory is primarily concerned with the formal analysis of collection of information to generate a social choice or preference, and Condorcet is credited with the basic building blocks of the theory; it was left to Kenneth Arrow in 1951 to revive it into its modern form.

[7] Sen (1999a, b), also Fleurbay and Mongin (2005).

[8] Malthus (1978), in the Penguin Classics edition, edited by Flew (1982).

[9] Swami Madhavananda (1950, 3rd edition). The *Brihadaranyaka Upanishad*: Sanskrit text, English translation. Durga Piture Lane, Calcutta, India: Advaita Ashrama.

Straight cut to the present, many of the political parties in India—both regional and national—are led by women. Even in national movement for independence led by the Congress party, there were many more women of influence, than in all global revolutionary movements put together. Sarojini Naidu, the first women president of the Indian National Congress was elected in 1925, fifty years earlier than the election of the first woman leader of a major political party (Margaret Thatcher in 1975). The second woman head of the Indian National Congress, Nellie Sengupta, was elected in 1933.Inequality between men and women is not just one affliction but a multitude of problems and has many faces.[10] Indeed, the absence of gender disparity in one dimension and the presence of the same in another sphere may coexist. For example, in Japan there is no gender bias in nutrition or healthcare or school education but the relatively better and leadership roles in business and administration are taken by men. While gender disparity has many dimensions, they are not necessarily independent; rather they may reinforce each other. For example, when women are deprived of decision-making power within the family, it leads to a deprivation of their effective agency,[11] and this can adversely affect their well-being.

10.3 Gender Disparity in India: Then and Now

The structure of inequality and gender disparity in India is the product of a complex history of involving the transformation of a premodern trifunctional society whose evolutionary trajectory was altered by its brush with the British colonizers. It is beyond the scope of this paper to delve into a speculation about how India's inequality regime might have evolved in the absence of the East India Company (1757–1858) and the direct administration of the British Empire (1858–1947). However, India's tryst with affirmative action and quota or reservation system is worth a mention. All available evidence suggests that the above policies that India had pursued since independence have significantly reduced inequalities between the old disadvantaged castes and the rest of the population- more, for example, than inequalities blacks and whites in the USA and much more than between blacks and whites in South Africa since the end of the apartheid.[12] Many studies have revealed that the measures adopted in India through parliamentary democratic procedures had the effect of bringing the lower classes into the electoral politics, and it is highly unlikely that such a result could have been achieved in any other fashion.[13] In 1993, a constitutional amendment required states that had not already done so to reserve a third of leadership posts in panchayats (village councils) for women. Research has shown that experiments

[10] Amartya Sen has addressed this issue in "Many Faces of Gender Inequality", *New Republic*, 17 Sept. 2001, and *Frontline*, Nov. 2001.

[11] Sen (1999a, b). In this book, he argues, rather convincingly, about the role of agency and that of well-being as how they are quite central to the process of development.

[12] Piketty (2020).

[13] Jensenius (2017).

with female-led panchayats have helped to reduce negative stereotypes of women, which may be the most convincing proof of the usefulness of affirmative action in overcoming longstanding prejudices.[14] The point of the above discussion is not to idealize the way India addressed the legacy of extreme inequalities, but to note that it is possible to draw any number of conclusions from India's experience.

Generally speaking, every country bears traces of its own particular political-ideological trajectory, whether it be social-democratic, post-communist, postcolonial, or petro-monarchical. Today's inequality regimes combine elements of modernity and archaism. Some institutions and discourses are new, while others reflect a return to old beliefs, mainly patriarchy. Most societies throughout history have known one form or another of male domination, especially with regard to political and economic power. In many Western countries, it was not until the 1960s and 1970s that married women were allowed to sign work contracts or open bank accounts without their husband's approval or that the law ceased to treat male and female adultery differently in divorce. The battle for women's suffrage was long and conflictual and is not over yet. Women were successful in New Zealand in 1893, in the UK in 1928,[15] in Turkey in 1930, in Brazil in 1932, in France in 1944, in India in 1950, in Switzerland in 1971 and in Saudi Arabia in 2015.

10.4 Board Diversity and Indian IT Companies

The law governing corporate leadership for registered companies is dealt under the Indian Companies Act, 2013.[16] In General, companies in India have two-tier[17] board system comprising of supervisory board, also known as executive board and management board. Executive board has persons who cannot be removed by general meeting, and they are non-rotational directors. Whereas the management board comprises of persons elected by the shareholders in general meeting and one-third of the total non-rotational directors is liable to retire by rotation at every annual general meeting.[18] Further, the board comprises of different categories[19] of directors to protect and represent the interest of various persons. Diversity and inclusiveness in board leadership had taken place centre stage discussion in many forums, and some relief was brought to incorporate the same by the Companies Act,2013. 2013 Act has brought remarkable changes to the Indian corporate law landscape and one such change is

[14] Beaman et al. (2009).

[15] In the UK, women's suffrage was granted in stages. Women over the age of 30 who met a property ownership condition obtained the right to vote in 1918; parity with men on the right to vote was instituted in 1928.

[16] Chapter XI of Indian Companies Act, 2013 Ss.149–169.

[17] Indian companies are predominantly controlled by promoters, who generally have proprietary interest in the affairs of the company. Most public listed companies had its origin in Joint Hindu Families and were later converted into Joint Stock Companies for varied reasons.

[18] Section 152 of the companies Act, 2013.

[19] Independent directors, small shareholders directors, nominee director, women director, etc.

the recognition of women directors. A significant characteristic of Indian companies is the vesting of ownership in the hands of the promoters,[20] similar to big family-owned corporate houses. The promoters influence appointments, decision-making and reporting. Section 149(3) read with companies' appointment of directors' rule mandate the appointment of at least one woman on the board of every public listed companies. Underlying reason behind the said provision is to initiate not only to ensure representation to women, but also initiate corporate policies and strategies that would be beneficial to the women stakeholders of corporation. Women constitute more than 40% of the total population in India, yet insufficient recognition was given by the Companies legislation to make women a forerunner in corporate leadership position.[21]

Management control of most public listed companies is by promoters, and promoters are permitted to hold 75% of the total paid-up equity shares. Because a significant characteristic of Indian companies is the vesting of ownership in the hands of the promoters and promoters influence appointments, decision-making and reporting. However, with specific reference to IT, which has been the fastest growing Indian industry since globalization, has not only given the Indian woman freedom and self -dependence, but it has also created a hitch for her in the world, whereby she interacts with people from diverse walks of life.[22]

According to NASSCOM survey on 150 IT companies, FY 2018 had over 34% women workforce (1.3 to 1.4 million). The term IT includes a diverse range of activities. The breakdown of the female representation in each sector was as follows: BPO, KPO and ITES (37.37%); internet business (37.37%); and software, hardware and information technology (33.33%).

Figure 10.1 shows that 46.2% of Indian IT firms have over 31% of women workforce. This has been a positive trend for women in IT in India. However, the same trend is not reflected at the highest levels tails off markedly. Figure 10.2 shows that most companies surveyed (88.5%) reported that less than 10% of their C-Suite is female. The number improves slightly for women in Senior Management—80% of companies have less than 10% women. The Middle Manager and Individual Contributor levels are fairly similar—around 65% companies have less than 10% women, and around 18% of them have between 11 and 20% women at this level. The higher the career level, the lower the participation of women. Though board diversity is quite impressive at the entry level but towards the higher order of the progression, it declines significantly.

The situation is alarming as board diversity is important to improve corporate governance as it promotes inclusion of the interest of all stakeholders to create purpose-driven corporation having sustainable objectives. Diversity of the board in terms of gender is advocated in many developed jurisdictions, to empower women as change maker and equal participant in the economic contributions to the society.

---

[20] Promoters of Indian Public listed companies are permitted to hold up to 65% of the total paid-up capital, this confers them a huge power in choosing the management of the company.

[21] Srinivasan and George (2013).

[22] Chakrabarty and Bhutia (2007).

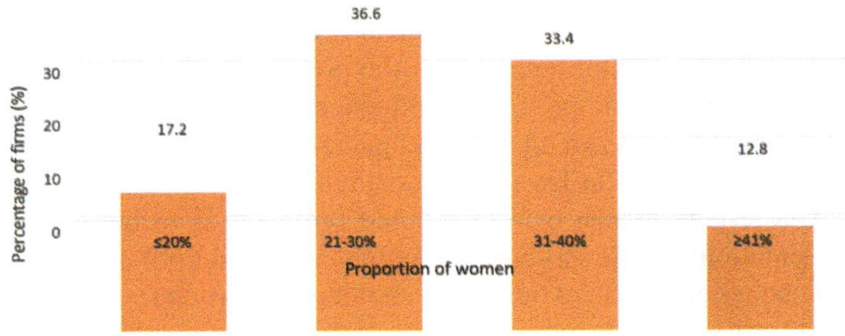

Fig. 10.1 Female employees in firms (as a percentage of firms' responses). *Source* NASSCOM IT-scorecard-India (2018), survey of 150 Indian IT companies

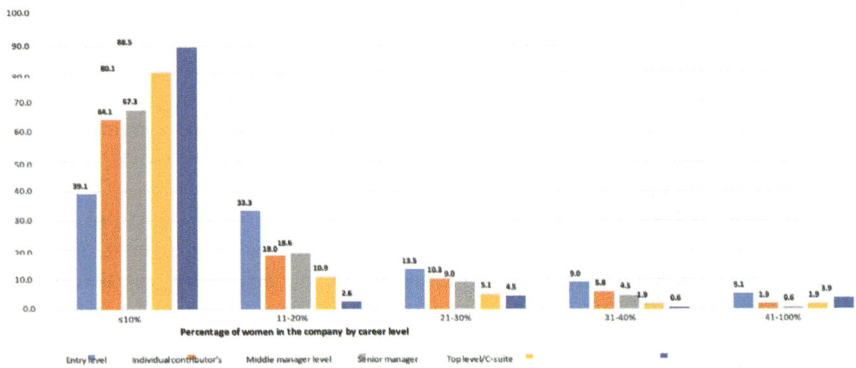

Fig. 10.2 Percentage of women employed at various career levels. *Source* NASSCOM IT-scorecard-India (2018), survey of 150 Indian IT companies

In this situation of pandemic, IT companies are playing crucial role as facilitators of business and other services to carry on. Technology is an enabler and without which, it's impossible for business entities to survive. Automation of industries and technological innovations are key not only for business to flourish but also for the society to improve. Board comprising of male and adequate number of female leaders will foster not only economic purpose but also that of social objective of its existence. Purpose-driven sustainable corporations are the pillars of economic development of every nation. Such corporations are possible not only if it has independent directors, but also directors who understand the concerns of its women employees, who are part of the workforce of industries in the recent times. Switzerland, USA, Singapore and many other developed jurisdictions have amended their laws to provide more opportunities to a greater number of women even to take up leadership role in companies.

Women leadership in IT can make a strong contribution to the economic wellbeing of the family and communities, poverty reduction and women's empowerment, thus contributing to the economic development of the nation. However, women leadership still continues to be negligible. Women often face gender-based challenges at their workplace. As per the World Bank's annual Doing Business report, the number of new limited liability companies in India as of 2018 is 12,39,42. Among this, it is estimated that presently women entrepreneurs comprise only about 14% of the total entrepreneurs. Cross-country data from the World Bank Entrepreneurship Snapshots find that India's rate of entrepreneurship is lower than its stage of development, and similar comparisons also highlight that India's gender ratio among entrepreneurs is lower than its peers. The Global gender gap report of 2020, states that only 13.8% of the Indian women are occupying corporate leadership positions. The worldwide statistics on women participation in managerial positions reveals that there is only 20–40% representation in about 48 countries, whereas the representation of women in senior managerial positions is much lesser.

Board leadership as well Chairmanship in Indian IT companies were found to be male preserve, with less than 1% of the position being held by women. While examining the board composition of 14 leading IT companies, we have found that the representation of women is too low compared to that of male leaders.

Figure 10.3 shows that only in Tech Mahindra Ltd and HCL infosystem Ltd have more than 2 women directors on Board, while in other companies its two or less than 2 in number. Majority directors in these companies are appointed not due to their technological competence but because of the promoter stake they hold in these companies. This is revealing that women are neither holding promoter stake nor leadership positions in the Indian IT companies, and therefore, they are unable to exercise influence by way of management control or by way of shareholding. Thus, women are unable to take up leadership positions in the Indian companies. While this study has found that in HCL and in Tech Mahindra, a total of 3 women are occupying the leadership position. The reason is that they belong to the promoter's family and their families have the shareholding control over them.

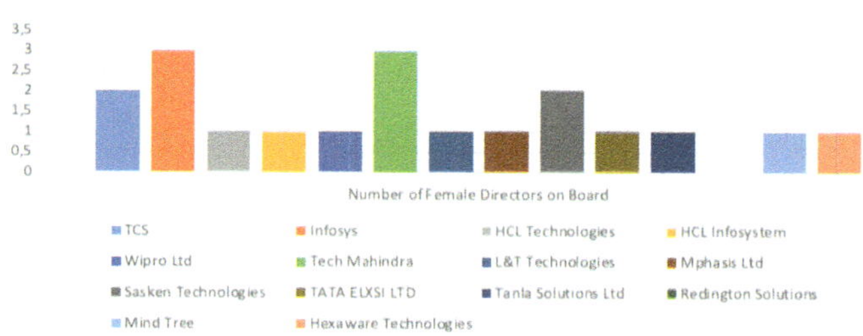

Fig. 10.3 Number of women directors at 15 Indian IT companies

Figure 10.4 shows that in terms of statutory mandate under Sect. 149[23] of Companies Act, 2013, above-listed IT companies have appointed women directors on their board, this shows that out of 15[24] statutory mandated directors in public listed companies only one has to be a woman. Legislature could have increased the number of women not only of the companies with higher turnover but also of non-public listed companies. The statutory criteria for prescribing women to be on the board are due to the reason that those are listed public companies or a company with the prescribed turnover. The gender diversity in corporate board is an unresolved issue in most jurisdictions.[25] In Fig. 10.3, HCL Technologies have a total of 20 directors out of which there is only one woman on board. Five companies out of the 14 used for this data have more than 10 directors on the board, but only one woman in these companies. We have identified that if the law had prescribed a higher number of women as a statutory mandate in the corporate board, then more representation of women would have been there on the boards. Therefore, from Fig. 10.3 we suggest that the legislative mindset of prescribing at least one woman should be reformed to at least one-third. This will bring women into the fore of business leadership as enablers of women empowerment. Practical reasons for not recognizing women leadership in corporate boards are unanswered. India has women leading risky business concerns like banking, insurance, pharmaceuticals as well as in the non-registered entrepreneurial activities and also in NGOs.

10.5 Women as Today's Leaders

Recently (Fig. 10.2), there has been a steady upward movement of women in the career ladder. Positive and steady growth has begun due to gender parity being one of the main factors of sustainability goals of many Indian IT companies. As India

[23] Section 149(1)(b) second proviso read with Companies (Appointment and Qualification of Directors) Rules 2014—The following class of companies shall appoint at least one-woman director-

(i) every listed company;

(ii) every other public company having -

(a) paid-up share capital of one hundred crore rupees or more; or

(b) turnover of three hundred crore rupees or more:

Provided that a company, which has been incorporated under the Act and is covered under provisions of second proviso to subsection (1) of Sect. 149 shall comply with such provisions within a period of six months from the date of its incorporation:

Provided further that any intermittent vacancy of a woman director shall be filled-up by the Board at the earliest but not later than immediate next Board meeting or three months from the date of such vacancy whichever is later.

Explanation.—For the purposes of this rule, it is hereby clarified that the paid-up share capital or turnover, as the case may be, as on the last date of latest audited financial statements shall be taken into account.

[24] This number can be increased by Passing special resolution.

[25] Balasubramanian (2010). Author has incorporated the data collected from USA, EU, UK and Germany as on 2008 and states that not more than 11.4% of women are on the board of corporations.

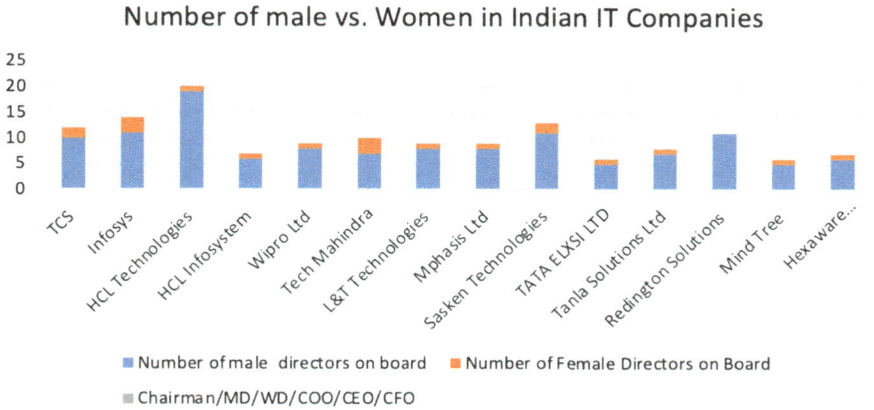

Fig. 10.4 Comparison of male and females leaders in Indian IT companies

continues to be the leader in IT outsourcing, its challenges are ever growing since then. Its unprecedented growth demands more skilled manpower and such a demand can only be filled by the women workforce provided there are conducive policies within corporations that support women at the workplace.[26] Indian IT companies that are interested in women's empowerment are either driven by corporate social responsibility mandate[27] or by business strategy. In both the cases, companies are equipped with more tools and support in the form of policies to achieve greater gender parity. Due to the increase in female employees in the IT sector and the increase of female participation in management, there have been some organizational support for the women in the form of talent management policies. Talent management policies are for attracting, developing, connecting and deploying the workforce and ensuring optimize employee performance. On the lines of gender parity, they aim to provide work/life balance, flex work programmes and peer support in order to make women part of the company's business strategy. Many of its work/life programmes are women-centric. But these practices are less seen in their use. Those who take advantage of these programmes are often the subject of unconscious and conscious biases and perceived as lacking commitment to their organization.[28] Hence, perceptions play an important role in women's job distribution regardless of what policies are in place to support the employees. Therefore, it is far from clear that talent management policies effectively addressed the particular needs of the women in the company reflecting the acute work–life balance challenges they face. This suggests that HR policies need to change fundamentally to address challenges related to promotion and

[26] Gallardo and Thunnissen (2016).

[27] Section 135 of the Companies Act, 2013 mandates CSR on companies with net worth of Rs. 500 crore or more or with turnover of Rs. 1000 crore or more or with net profit of Rs. 5 crore or more. Further as per the CSR Rules, the provisions of CSR are not only applicable to Indian companies, but also applicable to branch and project offices of a foreign company in India.

[28] Bourgeois and Traub (2020).

retention of female talents. Organizational policies on retention would help women who took a work break to come back to the workforce. This is more important in the Indian context reflecting the more acute nature of the work/family challenge and the greater tendency of women in India to drop out of the labour market in the child-rearing years compared to Europe. In addition, HR policies, work from home and a supportive manager could also be a great support. However, in IT companies this could depend on the approval of the clients of the Offshore Development Centres.

10.6 Women in the Labour Force and the Effect on Economic Growth and Productivity

The ongoing COVID-19 pandemic has drastically slowed down economic growth around the world, and an immediate measure to boost economic revitalization would be to make better use of women in the labour force. Since 2000, the Indian economy has grown at a great speed, recording the fastest growth of 8 per cent among G20 countries, but despite these economists have witnessed the phenomenon of "jobless growth", owing mainly to the lack of job creation by the services sector, in addition to the weak industrial and agricultural growth, the gradual mechanization of the formerly labour-intensive sectors, the global meltdown of 2008. It is disheartening to note that only 27.4% of women are currently employed as against 75.5% of men, and this translates to a gender gap of approximately 48.1% points,[29] pushing India to the fag end of the line, only behind Saudi Arabia, as far as female work participation rates are concerned, among G20 countries. According to a study by Lahoti and Swaminathan,[30] the women's employment rate (for 25–59 year olds) decreased by 25% points between 1983 and 2021. The reasons are manifold—the delayed entry of women into the labour market owing to their pursuit of higher education and extended training, an increase in the household median income, and also a growth story based on domestically driven demand where women do not participate much. It is further exacerbated by the deep-rooted gender inequality in the labour market where male dominance in political, economic, social, religious and cultural spheres is still dominant. Patriarchal norms and values not only prevent the entry but also the success of women in the labour market, and these norms are further legitimized and reinforced by the institutions of family, caste, community and religion. However, all is not lost as far as the relationship between gender and technology is concerned, in the Indian context. Emerging research points to the refreshing change of patriarchal norms, coupled with the changing economic scenario in encouraging women to study science, math and engineering enabling them to pursue careers in the information technology sector.[31] Interestingly, this is in stark contrast to the experience of female

[29] Government of India (2016a).

[30] Lahoti and Swaminathan (2016).

[31] Namrata (2015).

IT engineers in the USA, UK and much of Europe.[32] However, in spite of the spurt in the number of female IT engineers as well as the improvement of job avenues in the IT sector, gender segregation in the Indian IT labour market is still rampant.

As noted above, an increasing participation of women in the labour force augurs well in raising productivity, efficiency and strengthening corporate performance. Contemporary research points in the direction that the presence of women on corporate boards can enhance skills diversity and enable the board to improve the quality of the management. There is evidence to believe that exclusion of female executive can adversely affect corporate performance. This is particularly true in the case of Japan where "firms with more female outside directors exhibit higher performance. On average the market reacts positively to a firm's decision to bring a new female outside director on the board. Overall, the results show that female outside directors are beneficial to Japanese firms".[33] Further, research suggests that discrimination is costly and non-discrimination pays. In the USA, over the last 50 years, an increase in the number of women and African-Americans in highly skilled occupations resulted in a 20–40% productivity surge just by an improved matching of talent to job.[34] To continue, research suggests that female executives can better assess the productivity of female employees, evidenced by the finding that a given 25% of female employees when managed by a female CEO as against a former male CEO, can increase sales by an average of more than 3%.[35] Another interesting study finds that increasing the percentage of women in corporate leadership from zero to 30% translates to a 1% point increase in net profit margin, or about 15% increase in profitability.[36] Yet another study undertaken in the Indian context concludes that board gender diversity exerts a positive impact on the market performance of high-tech firms, whereas it negatively influences the performance of firms in the non-high-tech sectors.[37]

10.7 Comparative Position of Women on Board

Women corporate leaders are few and far between, both globally and in Asia, but Southeast Asia has an enviable record in gender diversity, especially in the prominence of women corporate leaders. The percentage of women board members was 20 in Thailand, more than 30% of executive officers were women in Malaysia, and in the Philippines, 95% of firms had at least one-woman executive officer. Japan is the only outlier in this respect with just 2% share of women corporate leaders. In New Zealand and India, the share of women board members was a paltry 6 per cent in 2006 but by 2017 it rose to a respectable 16% in both countries. In Europe, legal

[32] NSF (National Science Foundation) (2004), and Wajcman (2006).

[33] Tanaka (2019).

[34] Hsieh et al. (2019).

[35] Flabbi et al. (2019).

[36] Noland et al. (2016).

[37] Mukarram et al. (2018).

quotas were instituted with Norway leading the way- mandating that women presence be 40% in all the boards of publicly traded firms. As mentioned earlier, India too introduced similar laws in 2013, which resulted in a surge in the female share of board members between 2013 and 2015,[38] following it up with more stringent norms whereby the top 1000 listed firms were mandated to include "independent" female directors since April 2019. This could not have been made possible without the presence of a supportive educational system in the countries mentioned above, more so in the case of India. Indian families actively encouraged their women wards to pursue science and mathematics in school which was done with the intention of facilitating their entry into engineering courses later, which would in turn bring prestige of an IT job to the young woman and her family and further give the family an upper hand in marriage negotiations.[39] Armed with a lucrative job in the coveted IT industry with its vast networking capabilities and tools, the women's capacity to take decisions is enhanced, thereby further increasing their "agency" as discussed earlier.[40] Another factor which is worth the mention is the role of public policy, with enforceable laws as the starting point. Even in settings without explicit discrimination, careers of women diverge from that of their male counterparts during the child-rearing raising phase. Carefully drawn maternity and paternity policies, and child care avenues can shorten the absence of women from the labour force and can ease their re-entry. Quota systems, even though controversial in many countries, can come in handy and can produce desirable outcomes.

Thus, it can be easily said that, Indian IT sector is recognizing the growing influence of women power and has been gearing up to catalyse its growth and provide women with rewarding careers. Some of these policies are implemented as part of corporate social responsibility (CSR). CSR has been redefined as "the responsibility of enterprises for their impacts on society". These programmes were considered as heretical a few years ago and are now acknowledged as key tools for corporate productivity.[41] Thus, Indian IT industry has taken a lead in adopting best practices in HR to attract women employees and to enable IT as most preferred career choice. On the other hand, most of women workforce in India is highly concentrated in the low strata of employment such as coding and programming and only a minor percentage is in development of IT, which still needs a greater attention and efficient mechanisms in place.[42] The study has found the following factors contributing to gender disparity at leadership positions in Indian IT sector.

[38] Women and IT Scorecard-India, report by NASSCOM, 2017.

[39] Namrata (2012) and Varma (2011).

[40] Kelkar et al. (2002).

[41] Cramer and Pearce (1990).

[42] Shook and Sweet (2018).

10.8 Barriers to Women's Leadership

Despite strong commitments and policies in place, low levels of women's participation in the board room remain the norm. Structural and cultural barriers keep women from fully participating in the career growth. There exist external and subtler barriers that continue to thwart women's advancement in leadership. Though working women in general face challenges in managing multiple responsibilities, women particularly in the IT sector are more vulnerable towards following challenges.

i. **Social bias**: The bias or prejudice against women that they cannot perform as well as men in top-level corporate jobs have been the most important reason for boards to still remain male-dominated. Described as the "glass ceiling effect", it constitutes an invisible barrier that women face in the path to the top. Women often do not have mentors to guide them or role models to emulate. Men avoid mentoring women out of fear that their relationship will be considered personal, and there are simply not enough women at the top to mentor female board aspirants.[43]
ii. **Reconciling Work/Family Balance**: Women primarily face work overload and time constraints due to the twin role-played leading to stress. Working women cannot excel in careers without family support. The challenges include insufficient time for household responsibility and childcare, lack of support from the family which in extreme cases results in dropping out of the workforce due to family obligations.[44] Work/family reconciliation is now becoming a pivotal policy issue at the heart of gender equality in private and public life.[45] Legislative efforts do not currently have the momentum required to create worldwide impacts. Even CSR initiatives of corporations are generally chosen based on profitable and other priorities of the organization. Therefore, voluntary adoption of more gender-diverse boards is unlikely until circumstances make it beneficial. The state has also not acted as effective catalysts to make this a priority.[46]
iii. **Window dressing**: Another possible barrier is when promoters resort to the practice of window dressing, in order to comply with the mandate of Law.[47]Best lesson can be learnt from Norway. The Norwegian companies, in order to comply with the provisions of reservation promoted many women, with much less experience than their predecessors, to the position of the directors. However, this has a flip side too. Such a mandatory promotion of women would certainly make firms compromise the efficiency and would result in the reduction of the corporate

[43] Mishra (2016).

[44] Sam (2014).

[45] de Silva de Alwis (2013).

[46] Woryk (2011).

[47] Section 149(1) of the Indian Companies Act, 2013 coupled with Sect. 149(1) read with clause 49 of the SEBI Listing Agreement and Rule 3 of The Companies (Appointment and Qualification of Directors) Rules, 2014 provide that every listed company or public company with paid-up share capital of 100 crore rupees or turnover of 300 crore rupees or more shall appoint at least one-woman director.

productivity/profits. However, it is too early to reach any conclusion regarding this provision since the data available to establish this is very sparse.[48]

iv. **Part-time jobs**: Many women employees are employed in part-time, temporary, contract or other types of flexible work arrangements that lack job security. More than a quarter of women workers work part-time, making them 1.5 times more likely to be so employed than the average worker. While women are more likely to choose part-time work, they are also more likely to be stuck in part-time jobs against their will. These part-time jobs are mostly dead-end positions because part-time workers keep their jobs for shorter periods than full-time workers and often lack health and pension benefits.

v. **Lack of career advancement**: Advancement of jobs has not kept pace with the corresponding number of working women in Indian IT sector. The numbers of women graduating from IT universities is more than 50% and their entry into the industry has been quite great in the recent past. However, there is a bottleneck for women's entry in senior management levels due to career breaks and lack of training after their come back.

vi. **Government Policies and Regulations**: Though there are enough government policies in place to encourage women education and safety at work. Issues like long commute and work at odd hours continue to bother women. There is a need for timely and effective law enforcement action in case of harassment and emergencies. Efficient implementations of policies on work hours, late hours and maternity benefits must be ensured.

Based on the above analysis, the roadblocks faced by women in Indian IT industry can be presented in a fivefold classification such as professional, family, personal, social and legislative classes[49] as shown in Fig. 10.5.

10.9 Conclusions, Findings and Suggestions

The study has found that there is absolute imbalance in terms of gender-diversity on the board of Indian companies. Though the situation has begun to improve, particularly since 2011, the change has been negligible. Despite the amendment in Companies Act, 2013, new strategies and policies for the promotion of women employees to executive leadership positions including ensuring the disclosure of diversity of corporate boards should be considered as a prerequisite to list the shares on stock exchanges. Secondly, Companies Act, 2013 should prescribe higher number of mandatory appointments of women on board and to make it mandatory for women to be part of committees of board mandated under the Companies Act. Thirdly, it proposes that the B-Schools admission policies need to be revised to increase the intake of women candidates for management studies programme, so that the

[48] Vaibhavi and Soundarya (2015).

[49] Gupta, C. and KVS, R. R., Review of the Work-Life Balance Challenges of Women Professionals in Information Technology Industry.

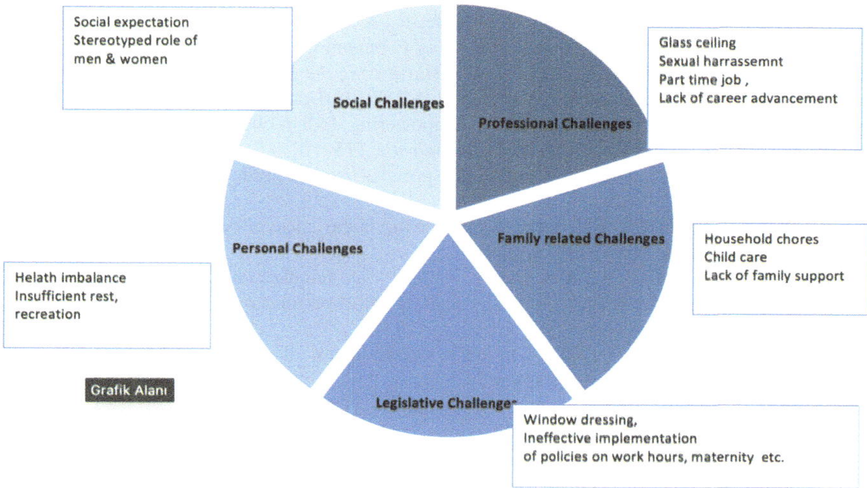

Fig. 10.5 Classification of challenges faced by women in IT industry in India

said women candidates would possess the adequate competency to govern corporate boards of Indian IT companies. A mandate should set the direction desired but should not be blindly followed, and competency should also guide the selection for the highest levels. Focus should be on the large number of women executives who are stuck at middle level positions for years and budgets should be allocated to reskill and retrain them for higher position.[50] Therefore, mere introduction of legislation is not sufficient for empowerment of women; companies should understand the benefits of gender diversity on their boards.[51] In this light, the role of Chairman and CEO is critical in making a strong business case for more women representation on boards.[52]

In the light of all of the above, it would be safe to surmise that the movement towards greater gender equality is somehow not a "natural" and irreversible process, and the need of the hour is to bring about concerted and proactive measures to bring about a desirable level of parity.

References

Arora, S., & Kumar, N. (2016). Women on boards: A gap analysis of India vis-a-vis World. *Chartered Secretary*.

Balasubramanian, N. (2020). *Corporate governance and stewardship*. TATA Mc Graw hill Publications.

50 Sundar (2012).

51 Arora and Kumar (2016).

52 Kishore (2016).

Beaman, L., Chattopadhyay, R., Duflo, E., Pande, R., & Topalova, P. (2009). Powerful women: Does exposure reduce bias? *Quarterly Journal of Economics*.

Bourgeois, T., & Traub, L. (2020). *Strategies for Harnessing the power of female talent: Changing the game*. https://cookross.com/wp-content/uploads/2014/01/changingthegame2012.pdf13

Chakrabarty, M., & Bhutia, Y. L. (2007). Empowering Indian women in the new wake of globalization. *The Indian Journal of Political Science*, 115–121.

Cramer, K., & Pearce, J. (1990). Work and family policies become productivity tools. *Management Review, 79*(11), 42.

de Silva de Alwis, R. (2013). Why women's leadership is the cause of our time. *UCLA Journal of International Law and Foreign Affairs, 18*, 87.

Flabbi, L., Macis, M., Moro, A., & Schivardi, F. (2019). Do female executives make a difference? The impact of female leadership on gender gaps and firm performance. *The Economic Journal, 129*(622).

Fleurbay, M., & Mongin, P. (2005).The news of the death of welfare economics is greatly exaggerated. *Social Choice and Welfare, 25*.

Flew, A. (1982). *An essay on the principle of population*. Penguin Books.

Gallardo, E., & Thunnissen, M. (2016). Standing on the shoulders of giants? A critical review of empirical talent management research. *Employee Relations, 38*(1), 31–56.

Government of India. (2016a). *Employment and Unemployment Survey*, Vols. 1, 2, and 4. Ministry of Labour and Employment Labour Bureau.

Hsieh, C.-T., Hurst, E., Jones, C. I., & Klenow, P. (2019). The allocation of talent and U.S. economic growth. *Econometrica, 87*(5).

Jensenius, F. (2017). *Social justice through inclusion: The consequences of electoral quotas in India*. Oxford University Press.

Kelkar, G., Shrestha, G., & Veena, N. (2002). IT industry and women's agency: Explorations in Bangalore and Delhi, India. *Gender, Technology and Development, 6*(1).

Kishore, K. (2016). Representation of women on boards of IT companies: An Indian story. *Journal of Management and Public Policy, 7*(2), 29–36.

Malthus, T. R. (1978). *Essay on the principle of population, as it affects the future improvement of society with remarks on the speculation of Mr. Godwin, M. Condorcet, and Other Writers*. J. Johnson.

Mishra, R. K. (2016). Diversity and board effectiveness: A case of India. *Journal of Modern Accounting and Auditing, 12*(3), 165–177.

Mukarram, S. S., Saeed, A., Hammoudeh, S., & Raziq, M. M. (2018). Women on Indian boards and market performance: A role-congruity theory perspective. *Asian Business and Management, 17*.

Namrata, G. (2012). Women undergraduates in engineering education in India: A study of growing women's participation. *Gender, Technology and Development, 16*(2).

Namrata, G. (2015). Rethinking the relationship between gender and technology: A study of the Indian example. *Work, Employment and Society, 29*(4).

Noland, M., Moran, T., & Kotschwar, B. R. (2016). *Is gender diversity profitable? Evidence from a global survey*. Working Paper No. 16–3. Peterson Institute for International Economics.

NSF (National Science Foundation). (2004). *Women, minorities, and persons with disabilities in science and engineering. NSF04–317*. Division of Science Resources Statistics.

Nussbaum, M. (1999). Women and equality: The capabilities approach. *International Labour Review*, 138(3).

Nussbaum, M. (2000). *Women and human development: The capabilities approach*. Cambridge University Press.

Piketty, T. (2020). *Capital and ideology*. The Belknap Press of Harvard University Press.

Prahlad, C. K. (2005). *The fortune at the bottom of the pyramid: Eradicating poverty through profits*. Prentice Hall.

Rahul Lahoti and Hema Swaminathan. (2016). Economic development and women's labor force participation in India. *Feminist Economics, 22*(2), 168–195.

Sen, A. (1999a). The possibility of social choice. *American Economic Review, 89.*
Sen, A. (1999b). *Development as freedom.* Knopf, and Oxford: Oxford University Press.
Sen, A. (2006). *The argumentative Indian: Writings on Indian culture, history and identity.* Penguin Books.
Sam, A. (2014). Work-life balance of women academicians–An empirical study.
Shook, E., & Sweet, J. (2018). *When she rises, we all rise: getting to equal 2018: Creating a culture where everyone thrives.*
Srinivasan, V., & Reji, G. (2013). *Building the women directorship pipeline in India: An exploratory study.* IIM-B working paper. Available at https://www.iimb.ac.in/sites/default/files/2018-07/WP_No_427_0.pdf
Sundar, P. (2012). Czarinas or girl fridays? Women in the corporate sector. *India International Centre Quarterly, 39*(3/4), 69–80.
Tanaka, T. (2019). Gender diversity on Japanese corporate boards. *Journal of the Japanese and International Economies, 51*, 19–31.
Vaibhavi, T., & Soundarya, V. L. (2015). Representation of women on the board of directors under the companies act, 2013. *Christ University Law Journal, 4*(1), 33–47.
Varma, R. (2011). Indian women and mathematics for computer science. *IEEE Technology and Society Magazine, 30*(1).
Wajcman, J. (2006). New connections: Social studies of science and technology and studies of work. *Work, Employment and Society, 20*(4).
Woryk, M. (2011). Women in corporate governance: A Cinderella's story. *University of Dayton Law Review, 37*, 21.

Index

M. A. Naser (ed.), *New Approaches to CSR, Sustainability and Accountability, Volume IV*, Accounting, Finance, Sustainability, Governance & Fraud: Theory and Application, https://doi.org/10.1007/978-981-16-9499-8

The manufacturer's authorised representative in the EU is Springer Nature Customer Service Centre GmbH, Europaplatz 3, 69115 Heidelberg, Germany. If you have any concerns regarding our products, please contact ProductSafety@springernature.com

Printed and bound by CPI Group (UK) Ltd, Croydon, CR0 4YY
30/09/2026
02225242-0001